# THE STORY OF
# SCIENCE
## FROM ANTIQUITY TO THE PRESENT

© 2010 Tandem Verlag GmbH
h.f.ullmann is an imprint of Tandem Verlag GmbH


Editors: Ritu Malhotra & Gaurav Dikshit
Design: Supriya Sahai & Baishakhee Sengupta
DTP: Neeraj Nath, Ajmal Khan
Arrangement: e.fritz, berlin06
Project coordination: Daniel Fischer, Ulrike Reihn-Hamburger
Cover design: Yvonne Schmitz
Arrangement cover: e.fritz, berlin06
Overall responsibility for production: h.f.ullmann publishing, Potsdam, Germany


Printed in China


ISBN 978-3-8331-5248-1


10 9 8 7 6 5 4 3 2 1
X IX VIII VII VI V IV III II I


If you would like to be informed about forthcoming h.f.ullmann titles,
you can request our newsletter by visiting our website (www.ullmann-publishing.com)
or by emailing us at: newsletter@ullmann-publishing.com.
h.f.ullmann, Birkenstraße 10, 14469 Potsdam

R.R. SUBRAMANYAM
SHOBHIT MAHAJAN
ARCHANA MADHUKAR
SUMAN SAHAY
G.S. ROONWAL

# THE STORY OF SCIENCE

## FROM ANTIQUITY TO THE PRESENT

h.f.ullmann

# CONTENTS

# BIOLOGY 113

# GEOLOGY 167

# PHYSICS

* PHYSICS IN THE ANCIENT WORLD

* FROM MIDDLE AGES TO RENAISSANCE

*SCIENTIFIC REVOLUTION

* MODERN PHYSICS

# Chapter 1
# PHYSICS IN THE ANCIENT WORLD

**HERACLITUS AND DEMOCRITUS,** Fresco, Donato Bramante, c. 1500, originally in the "Sala dei Baroni" of Palazzo Panigarola, Accademia di Belle Arti di Brera, Milan, Italy

The ancient Greek philosophers Heraclitus (c. 540-470 BC) and Democritus were the models of contrasting temperaments for Renaissance artists: the former being a melancholy philosopher and the latter an optimist with a laughing countenance. While Heraclitus is best known for his cosmology where fire is the essence of all things in the universe, Democritus is a central figure in the development of the atomic theory.

**FACING PAGE:** Historical artwork of Archimedes using a lever to move a ship. Archimedes claimed, "Give me a place to stand and rest my lever on, and I can move the earth." According to legend, when asked to move a ship to prove his claim, Archimedes moved the ship easily by using a compound pulley system.

The science of physics is perhaps as old as the evolution of man. About two million years ago, homo sapiens, not aware of the laws of friction, discovered fire by striking flints together. Fire was a prime energy source, eventually leading to the fashioning of metals into tools that created implements for plowing in agriculture.

Starting in prehistoric times, the story of physics spans across several ancient civilizations—Indian, Chinese, Babylonian, Mesopotamian, Egyptian, Greek, and Roman. The Judaic, Islamic, and Christian civilizations made their contributions to natural philosophy, which later developed as physics. The influence of the Greeks, in particular, has been well documented. From 1500 BC to AD 1000 phenomenal advances took place in the field of astronomy and mathematics. This laid the foundation for physics as we know it today.

## PHYSICS IN ANCIENT CIVILIZATIONS

The earliest of the ancient civilizations, Sumer (and later Babylon), emerged in the Mesopotamian region around 3500 BC. The Sumerians developed the number system and basic arithmetic. Around the same time, the ancient Egyptian civilization was growing in the Nile Delta. The Egyptians had a strong base in geometry—as can be seen in the precision of their pyramids—and developed an early calendar that had 365 days in a year.

All these developments had an influence on the thinkers in the ancient Greece in the Aegean Sea region. Here people believed that there were gods for the sky, the earth, thunder, oceans, and fire. Such culture inhibited the emergence of rational scientific thought. However, in the first millennium BC, there was an intellectual awakening in Ionia, Samos, and other Greek colonies in the islands of the Aegean Sea. Ionia was the breeding ground of scientific thought, perhaps because it was in the eastern Mediterranean where the diverse civilizations of Africa, Asia, and Europe interspersed and influenced each other. In the 6th century BC, the Ionians, having rejected the superstition prevalent at that time, were aware that the universe exhibited an order and that nature was not disorderly but obeyed certain rules.

Thales (c. 625-546 BC), a philosopher from Miletus (an Ionian city), pioneered scientific thought in the ancient world. He somewhat bridged the gap between superstition and science by

## GREECE

**3000-1100 BC** Bronze age. Early Aegean culture: Minoan and Mycenean civilizations.

**1100-800 BC** Dark age. First Greek migration to west coast of Asia minor.

**800-500 BC** Archaic period. Greeks search for new land for agriculture. City-states, that function as political units or polis, are formed.

**600 BC** Thales of Miletus states water is the essence of all matter.

**627 BC** Anaximander proposes that the origin of cosmos is not a supernatural phenomena.

**500-330 BC** Classical period. Greeks defeat Persian invaders. Athens becomes famous in the Mediterranean. The Peloponnesian wars. Alexander the Great's rule begins.

**480 BC** Anaxagorus discovers the cause of solar eclipses.

**425 BC** Democritus develops Leucippus' theory that all matter is made of small indivisible particles called atoms.

**335 BC** Aristotle rejects the atomic theory and presents his theory of five elements.

**280 BC** Aristarchus of Samos proposes that the earth revolves around the sun.

**240 BC** Archimedes discovers the principle of buoyancy.

**235 BC** Eratosthenes measures the circumference of the earth.

**330-30 BC** Hellenistic age. Greece becomes part of the Roman empire.

**270 BC** Ctesibius of Alexandria improves the clepsydra.

**130 BC** Hipparchus makes accurate predictions of the precession of the equinoxes.

**AD 130** Ptolemy produces the *Almagest*, presenting his geocentric model of the universe.

---

trying to understand the nature of the physical world through research and not mythology. He discovered the concept of static electricity by showing that a piece of amber, when rubbed with cat fur, could attract feathers. At a time when it was hard to understand the cycle of solar eclipses, Thales predicted a solar eclipse accurately. No wonder he was the first of the Seven Wise Men of Greece.

Around 450 BC, Empedocles (c. 490-430 BC), also an Ionian, showed by experiment that air existed and was not empty space. He used a clepsydra, a vessel with a hole in the bottom and one in the top. Placing the bottom hole of the vessel under water, Empedocles observed that the vessel filled up with water. If, however, he put his finger over the top hole, the water did not enter the hole at the bottom. Empedocles concluded that the air in the container prevented the water from entering. It is to the credit of Empedocles that he was able to prove his theory through the empirical evidence of experiment.

The Ionian influence and experimental methods spread to Greece, Italy, and Sicily (where Archimedes was born). In the 5th century BC, Anaxagoras (c. 490-428 BC) and Leucippus (mid-5th century BC) proposed that all matter was made of infinitesimally small particles. This idea was developed by Democritus (c. 460-370 BC), who came from a relatively unknown place called Abdera in northern Greece. Famous for his immense contributions to the development of science, Democritus was the first to point out that a large number of worlds had formed spontaneously from matter spread far and wide in the space. He thought such worlds could occasionally collide. These were far-reaching conclusions that were eventually

### THALES

Thales of Miletus is the first of the legendary Seven Wise Men, who introduced mathematical and astronomical sciences to Greece. It is believed that he predicted a solar eclipse on May 28, 585 BC, thereby stopping the battle between the Lydian Alyattes and the Median Cyaxares. He is credited with five theorems of elementary geometry and using geometry to measure the pyramids of Egypt.

# ANCIENT ASTRONOMY

Astronomy originated in Egypt and Mesopotamia. Egyptians developed a fairly accurate calendar, allowing astronomers to record their observations. Working with a highly developed system of observation and mathematical calculation, Babylonians made accurate predictions of astronomical phenomena, such as the sighting of the new moon.

In the 5th century BC, the Pythagoreans developed the first physical model of the solar system. Eudoxus of Cnidus (c. 395-342 BC) developed a theory of homocentric spheres in the 4th century BC, which was modified by his student Callippus. Around that time, Heracleides Ponticus (c. 390-322 BC) suggested that some planets revolve around the sun—which in turn revolves around the earth—and that the earth rotates on its axis. This theory was developed by Aristarchus of Samos (c. 310-230 BC) in the 3rd century BC, who put forth the earliest heliocentric theory.

In his theory of motion of objects, Aristotle proposed a geocentric theory of cosmology, where earth is at the center, surrounded by planets and stars in concentric spheres. This remained the dominant theory in astronomy till it was overthrown by Copernicus in the 16th century.

Eratosthenes of Alexandria (c. 276 BC-AD 194), a Greek astronomer, deduced that the earth was curved and calculated its circumference. This discovery made sea voyages possible.

Hipparchus (fl. 130 BC) developed trigonometric tables and rigorously applied geometry to the study of astronomy. His most significant contribution was the prediction of the precession of the equinoxes. Claudius Ptolemaeus or Ptolemy (c. AD 85-168), a Greek philosopher, took over from Hipparchus and perpetuated the idea that the world has the earth as its center. Ptolemy's detailed astronomical observations were recorded in his treatise the *Almagest*.

**HIPPARCHUS,** Line engraving, 19th century

Greek astronomer Hipparchus, who completed the first star catalogue, measured the distance of the sun and moon from the earth, and invented trigonometry, at the observatory in Alexandria.

**PTOLEMAIC SYSTEM,** From "A Celestial Atlas, or The Harmony of the Universe," *Atlas Coelestis Seu Harmonia Macrocosmica*, 1660-1661

**THE EDUCATION OF ALEXANDER THE GREAT BY ARISTOTLE,**
A Book by L. Figuier, French School

be described mathematically. Rejecting the theory of the atom, he proposed that all matter is made of five elements: fire, earth, air, water, and ether. He considered the cosmos to be divided in two realms—celestial and terrestrial—and studied the motion of objects in these realms. His observations on the nature of motion were an important step in the evolution of physics.

Although he formalized logic, Aristotle considered that mathematics could not be applied to the study of nature. Archimedes of Syracuse (c. 290-80 BC), the first mathematical physicist, creatively applied mathematics to the solution of many physical problems. He was thus able to derive the mathematical principle of the lever, create elaborate pulley systems, establish the concept of the center of the gravity, and deal with problems of the equilibrium of floating objects. He developed the physics of hydrostatics in the 3rd century BC, which eventually became hydrodynamics under Euler and Lagrange in the 18th century. There is a famous story as to how Archimedes solved the Greek emperor's dilemma about what was the ratio between gold and silver in his crown. While ruminating this problem in his bathtub, he suddenly jumped out naked, screaming "Eureka" ("I found it"). He was able to figure out the ratio of gold to silver by measuring the

developed many centuries later by Laplace in France. More significantly, Democritus is credited with coining the word "atom", which means "unable to be cut" in Greek. He laid the foundations for the atomic theory of matter, which was developed by the English chemist John Dalton in the 19th century. Elementary particles, which are a part and parcel of modern physics, do not bear out Democritus' theory that the atom is indivisible.

In the 4th century BC, during the reign of Alexander the Great, Plato, Socrates, and Aristotle—the three great philosophers of ancient Greece—formed the conceptual framework for the development of natural sciences. Socrates taught Plato, who in turn was the mentor of Aristotle (c. 384-322 BC). Unlike the moral philosophy of Socrates and Plato, Aristotle's natural philosophy was based on observation, though not so much on experimentation. He believed that observation of the natural phenomena could lead to the knowledge of natural laws causing these phenomena. According to him, these laws were divine in nature and could not

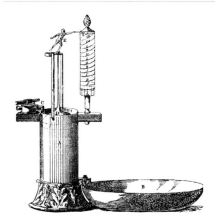

**CTESIBIUS' CLEPSYDRA,**
Engraving of a reproduction

The first water clock was developed in antiquity, perhaps by the Egyptians. Ctesibius of Alexandria improved the clepsydra to include a cylinder into which water dripped from a reservoir, raising the float that held a pointer to mark time.

**ARCHIMEDES,** Planning the defense of Syracuse, Engraving, 1740;
**Below:** Showing the use of reflecting mirrors to set fire to enemy ships.

Archimedes discovered the relation between the surface and volume of a sphere and its circumscribing cyclinder. Based on this, he formulated a hydrostatic principle (known as Archimedes' principle). He invented the Archimedes screw—a hydraulic device that raised water from a lower to a higher level—and also the catapult, the lever, the compound pulley, and the burning mirror (a system of mirrors that burned the enemy ships by reflecting the sun's rays).

amount of water displaced by the crown when immersed in water. Hence, the principle of Archimedes was evolved, which states the amount of water displaced by a body is equal to the magnitude of the upthrust or the buoyant force. The buoyant force on a body floating in a liquid, like a ship on water, is equal to the weight of the floating object and opposite in direction, so the ship neither rises nor sinks. This is the principle of floatation or buoyancy. This explains why iron sinks in water while a ship made of iron floats at sea. Archimedes is also credited with a number of notable inventions.

From 300 BC, after the death of Alexander the Great, the Roman empire grew in power and lasted till the 5th century AD. This empire extended from the Middle East to the British Isles. The Roman culture was not known for conceptualizing new ideas in science, but saw several innovations in the technological applications of science. The Romans were famous for their civil engineering achievements, such as aqueducts, dams, roads, bridges, and amphitheatres.

Ctesibius of Alexandria (*fl. c.* 270 BC) invented several compressed air devices, the most famous being hydraulis, a water organ in which air was pushed through organ pipes by the weight of water rather than falling lead weights. He also improved the clepsydra, or water clock, which kept the most accurate time till the Dutch physicist Christiaan Huygens

# PHYSICS IN ANCIENT INDIA AND CHINA

The earliest atomic theories are more than 2,500 years old. In the 6th century BC, Kanada, a Hindu philosopher, developed the atomic theory, where the atoms, the primordial stuff of the world, unite to form the gross bodies according to the divine will. Another Indian philosopher, Pakudha Katyayana, also propounded the ideas about the atomic constitution of the material world. Almost every school of philosophy in India—Hindu, Buddhist or Jain—supported the theory of elementary particles, which was expanded by later philosophers to the theory of dyads and triads, and the molecular theory of matter. The Indians were the first to suggest that light and sound traveled in waves. Mimasakas suggested that light comprised small particles—now known as photons. The earliest theories of reflection and refraction also originated in India.

Indian astronomy was quite advanced and made significant contribution to astronomy. The earth was assumed to be the center of the universe around which revolved the seven *grahas* (planets)—the sun, the moon, mercury, venus, jupiter, mars, and saturn. The stars, referred to as the *nakshatras*, provided the backdrop of planetary motions.

From the 2nd century BC to the 1st century AD, Babylonian and pre-Ptolemic astronomy and astrology influenced India. Indeed, India

**THE FIRST COMPASS,** Invented in China

became an active heir to a science that fell into decline in the West in the 5th century AD. Astronomical science that evolved in India from Greek sources is known as Siddhantic astronomy. Its most important source is the compendium of various astronomical systems written by Varahamihira (AD 505-587) in the 6th century AD. The mathematically rigorous Siddhantic astronomy began in AD 499 with the treatise *Aryabhatiya*, written by Aryabhata I (AD 476-550), the pioneer of scientific astronomy in India. He proposed that the apparent diurnal rotation of the heavens is due to the diurnal rotation of the earth on its own axis. After Aryabhata, Indian astronomers discovered the precession of the equinoxes (relating to the axis of the earth) and solstices (the apparent migration of the sun that causes

the seasons). There is evidence to show that Indian astronomers relied on their own observations and owed nothing to the Greeks. In fact, they had more correct values for the precessional rate than Ptolemy.

It was in the 14th century that observational astronomy came into its own. Indian astronomers since Aryabhatta I occupied themselves with the calculation of geocentric planetary orbits. Bhaskara I (*fl.* AD 629) and Bhaskara II (1114-1185) were the celebrated astronomers who developed Aryabhata's work further.

Buddhism came to China during the second half of the first century. The contact with India brought the knowledge of Indian mathematics and astronomy to China.

Zhang Sui (AD 683-727), also known as Jingxian and Yixing, was one of the greatest astronomers in China, who constructed new astronomical instruments in collaboration with Liang Lingzam in AD 724: an elliptically mounted celestial latitude ring called *huengdu youyi* and a water driver rotating armillary sphere *shuiyun hunyi*.

Chinese astronomers sighted the Crab Nebula in 1045. The high point in Chinese astronomy came around 1280 with the work of Kuo Shou-Ching (1213-1316), who introduced improved astronomical instruments and mathematical techniques for computation.

**HERO'S AEOLIPILE**

Hero of Alexandria invented the aeolipile, the first steam-powered engine. Like other machines of the time, this was just a toy and not put to any practical use.

**THE GREAT LIBRARY OF ALEXANDRIA**

Founded by Alexander the Great, built and enlarged by Ptolemy I (Alexander's successor), the Great Library of Alexandria was established around 300 BC. The library's lofty goal was to collect half a million scrolls, and Ptolemy and his son took serious steps to accomplish it. They acquired as many as 700,000 manuscripts from Mesopotamia, Babylon, Greece, and India. The library was a repository of knowledge and drew scholars from various parts of the world. It was the breeding ground of the scientific method and other new concepts. But fires and depredations during the Roman period gradually destroyed the library. In the words of a scholar, science was sent back a thousand years. Whatever little remained provided the basis for further development of science well into the 16th century.

invented the pendulum clock in 17th century. Ctes bius is the founder of the engineering tradition in Rome that flourished in the works of Philo of Byzantium and Hero of Alexandria.

Philo (*fl.* 260-180 BC) studied with Ctesibius and produced a vast body of works. The compendium of his works, *Mechanical Collection*, covers all aspects of mechanics and engineering. His inventions include the chain pump, the air pump, the piston pump, a treadmill-driven waterwheel, and other machines.

Hero of Alexandria (*fl.* 62 AD) invented a number of engineering devices, the most significant being the aeolipile, the first steam-powered engine. It is the first known device to transform steam into rotary motion, consisting of a hollow sphere—mounted by an axial shaft—and a pair of hollow tubes that caused the sphere to rotate by providing steam from a cauldron. Based on the work of Archimedes, Hero presented many principles of physics, including a theory of motion, a theory of the balance, methods of lifting and transporting heavy objects with mechanical devices, and the way of calculating the center of gravity for various simple shapes.

Other than this, the Romans were practical people who could not appreciate the scientific legacy they had inherited from Greece. Lacking the spirit of intellectual research, the Romans drove the surviving Greek scientists and philosophers to the East, where the Muslim rulers of a growing Islamic empire appreciated Greek scientific texts and preserved them in Arabian translations.

# Chapter 2
# FROM MIDDLE AGES TO RENAISSANCE

semblant figura. mas se les cors se
accident la dita varietat appar en
la descricha figura. III——III

item luna es de humors en son
crepshemet mlaphaniuacien son

**MEDIEVAL MANUSCRIPT SHOWING PHASES OF MOON,** *Le Propietaire des Choses* by Barthelemy de Glandville, 14th century

Astronomy was important in the medieval Islamic world to predict the first appearance of the crescent moon, which marks the beginning of the holy month of Ramadan.

**FACING PAGE:** Galata Observatory, founded in 1557 by Sultan Suleyman, From the *Sehinsahname of Murad III, c.* 1581

# MEDIEVAL ISLAMIC PHYSICS

In the 7th century, while Europe languished in the Dark Ages, the torch of ancient learning passed into Muslim hands. They kept it alight by preserving Greek scientific texts in Arabic translations. The new Abbasid dynasty, which took over the caliphate in 750 and founded Baghdad as the capital in 762, sponsored translations of scientific works of antiquity. By the 9th century, scientific works of the great Greek thinkers, such as Aristotle, Euclid, Ptolemy, Archimedes, and Apollonius were translated into Arabic, allowing the restoration of Greek science in the West in the 12th and 13th centuries.

Based on the Aristotelian framework, mechanics emerged as a popular field of study with the Arabs. Thabit ibn Qurra (*c.* 836-901) was one of the earliest Islamic scholars of mechanics. In *Kitab fi'l-qarastun* (*The Book on the Beam Balance*), he proved the principle of the equilibrium of levers. He also proposed a theory of motion, stating that both the upward and downward motions are caused by weight and that the order of the

universe is a result of two competing attractions. Mechanics formed a significant part of the first encyclopedia of science, *Mafatih al-'Ulum* (*Key to the Sciences*), compiled between 975 and 997 by the Persian scholar and statesman Al-Khwarizmi (10th century), who drew inspiration from Hero of Alexandria.

The Arab writers on mechanics pondered on the question of weight. The most comprehensive work in this field during the Middle Ages was *Kitab Mizan al-Hikma* (*The Book of the Balance of Wisdom*), written in 1121 by Abd al-Rahman al-Khazini (*fl. c.* 1115-1130). He presented a history of statics and hydrostatics, with commentaries on the works of his predecessors, including Euclid, Archimedes, Al-Biruni, Al-Razi, and Omar al-Khayyam. He discussed Greek and Arabic theories of the center of gravity and drew attention to the fact that the Greeks failed to distinguish between force, mass, and weight. However, like the Greeks, he considered gravitation as a universal force. Newton was to codify this concept quantitatively through the calculus in 18th-century England. Al-Khazini elaborated that gravitation as a force attracts all bodies toward the center of the earth.

His work shows awareness of the weight of the air and the decrease in its density with altitude.

The rest of the work was concerned with hydrostatics, especially Archimedes' principle of flotation. To determine the specific weight of a specimen, its weight has to be known in air and water, and the volume of air and water displaced by the specimen. Al-Khazini used a hydrostatic balance for this purpose, which was more accurate than the balances of his predecessors. He determined the specific weight of 50 substances, including 9 metals, 10 precious stones, and 18 liquids. He was the first to give the hypothesis that the gravity of a body varies with its distance from the center of the earth, a fact that Newton was to enunciate six centuries later.

## ASTRONOMY IN THE MEDIEVAL WORLD

Ptolemy's *Almagest*, the apex of ancient astronomy, formed the basis of Islamic astronomical studies in the Middle Ages. Astronomical research was greatly supported by the Abbasid caliph al-Mamun (786-833), who came to power in 813 and founded the House of Wisdom—an academy that vigorously acquired and translated Greek scientific texts and published new research. Al-Khwarizmi (*c.* 780-850) was a mathematical astronomer at the House of Wisdom, who compiled a set of

**PTOLEMY HOLDING UP A QUADRANT TO MEASURE THE ANGLE OF ELEVATION OF THE MOON,** *Margarita Philosophica* by Gregor Reisch, 1508

The Ptolemaic system, documented in his *Almagest*, dominated astronomy for over a thousand years. Arabic astronomers, building upon Ptolemy's work, significantly modified his theory, to the point of anticipating Copernican heliocentric theory.

**750** Beginning of the Abbasid era.

**830** House of Wisdom is established in Baghdad, promoting translations of Greek scientific works in Arabic.

**1011** Ibn al-Haytham writes the seven-volume *Book of Optics*.

**1100s** Founding of cathedral schools, and Universities of Bologna, Paris, and Oxford.

**1160** Ibn Rushd challenged the Ptolemaic system and started the Andalusian revolt.

**1259** Maragah Observatory is built in modern-day Iran.

**1260s** Al-Tusi introduces the Tusi couple, allowing Islamic scholars to modify Ptolemaic models of planetary motion.

**1277** Étienne Tempier, Bishop of Paris, issues condemnation on the Aristotelian philosophy.

**1305-1375** Ibn al-Shatir develops a planetary theory similar to the Copernicus model.

**1355** Hundred Years' War.

**1355** Jean Buridan gives the theory of impetus.

**1378-1417** The Great Western Schism.

**1400s** Renaissance in Italy.

**1425** The first printing press in Europe.

**1453** Fall of Constantinople to Turks.

**1480-1490** Leonardo da Vinci describes parachute, capillary action, flying machines, pendulum clocks, and compares reflection of light to reflection of sound waves.

# OPTICS

Optics rose as a scientific discipline in the Graceo-Roman world with Euclid's *Optika* (*c.* 300 BC), which gave the first theory of vision, stating that eyes emit rays of light in the shape of a cone, with its vertex at the center of the eyes and the base on the surface of the object being perceived. This optical "emission" theory opposed the earlier "intromission" theories, like the one suggested in Aristotle's *De anima* (*Tract of the Soul*). Euclid's theory of vision was based on the works of Hero of Alexandria and Ptolemy, and provided the ground for research in the Islamic world.

Al-Kindi (*c.* 801–873) was the first Islamic scholar of optics, who studied the propagation of light and the formation of shadows. He gave a description of the principle of radiation and developed the theory of emission of light. Based on Euclid's *Optika*, Al-Kindi's *De causis diversitatum aspectus* discusses the rectilinear propagation of light, the theory of vision whereby the eye emits three-dimensional rays to illuminate the object seen, and a theory of mirrors. In 984, Ibn Sahl (*c.* 940–1000) gave a law of refraction in his treatise *On Burning Mirrors and Lenses*, a study of how curved mirrors and lenses bend light.

**HISTORICAL ARTWORK OF VARIOUS OPTICAL EFFECTS,** From *Optica* by Polish physicist Witelo (13th century), 1535

Ibn al-Haytham (*c.* 965–1040) initiated a revolution in optics when he rejected the ancient theory that vision occurs by the emission of light rays from the eye in his *Kitab al-manazir* (*Book of Optics*). He promoted the intromission theory—vision occurs when eyes receive the light rays reflected from objects—and proved it by experimental demonstration. Ibn al-Haytham also investigated the properties of reflection from a number of surfaces with different profiles, and established the law of reflection as we know it today. He also dealt with the phenomenon of refraction of light and showed a correlation between the angle of incidence and the angle of refraction. However, he failed to discover the law established by Snel—that the ratio of the sine of the angle of incidence to the sine of the angle of refraction gives the

refractive index. These laws were eventually codified into the new dimension of physical optics developed much later by Newton and Huygens in Europe.

Qutb al-Din al-Shirazi (1236-1311) and his student Kamal al-Din al-Farisi (1260-1320) revised Ibn Al-Haytham's hypothesis of the rainbow in *Kitab Tanqih al-Manazir* (*The Revision of Optics*). The last major Arabic work in optics was the three-volume *Book of the Light of the Pupil of Vision and the Light of the Truth of the Sights*, written by Taqi al-Din (1526-1585) in 1574. The book contains experimental studies on vision, reflection, and refraction. The knowledge of all these developments in the Islamic empire eventually spread to medieval Europe and set the ground for modern physical optics.

**ARABIC BRASS ASTROLABE**

One of the most important scientific instrument of the Islamic world, the astrolabe consists of circles marked with angular measurements. By aligning the astrolabe with the horizon, the angular heights (altitudes) and positions (azimuths) of stars in the sky could be measured by astronomers. It was also used to calculate time in the Middle Ages.

century) and then by Gerard of Cremona (*c.* 1114–1187), the latter work providing Dante with the astronomical knowledge used in his *La Vita Nuova.*

Several religious problems in Islam gave an impetus to the development of spherical trigonometry in the study of astronomical phenomena. One, as Ramadan and other Islamic months begin when the thin crescent moon is first sighted, it was important to predict the visibility of the crescent moon. Two, as Muslims need to face toward Mecca during their prayers, they needed to determine the direction of the holy city from all locations and the time of the prayers from celestial bodies. The Islamic astronomers discovered new trigonometric functions to solve these spherical problems. This was made easier by

astronomical tables in his *Zij al-Sindh*, based on a variety of Hindu and Greek sources. The work contains tables for the movements of the sun, the moon, and the five planets known at the time. Al-Khwarizmi changed the way astronomical calculations were made. He invented the first quadrants and mural instruments, including the sine quadrants and the Quadrans Vetu, the first horary quadrant for specific latitudes. The medieval theory of the trepidation of the equinoxes is attributed to Thabit ibn Qurra, who translated many scientific treatises at the House of Wisdom, including a commentary on the *Almagest.*

In 850, Al-Farghani (9th century) wrote the *Jawani* (*Elements*), which helped to spread the basic and non-mathematical parts of Ptolemy's geocentric theory. Working with Ptolemy's theory and value of the precession, he reached the conclusion that it is not only applicable to the stars but also the planets. He determined the circumference of the earth to be 6,500 miles (10,460 km). The book was widely circulated through the Muslim world, and translated into Latin twice in the 12th century—once by John of Seville (*fl.* first half of the 12th

**DURER'S CELESTIAL GLOBE,** From *The Complete Woodcuts of Albrecht Durer* by Willi Kurth, 1936

Albrecht Durer (1471–1528) produced the first printed star charts in 1515. This depiction of the northern sky was prepared by him in conjunction with the astronomer Stabius. The constellation figures are depicted in a classical style. Four astronomers are shown in the corners: Aratus Cilis, Ptolomaeus Aegyptius, Maniliusromanus, and Azophi Arabus (Al-Sufi).

**LEO THE LION,** From *Le Livres des Etoiles* (*The Book of Stars*), Sicily, Italy, 13th century, Arsenal Library, Paris, France

The spots show the position of the stars that make up this constellation, one of the 12 star signs of the zodiac. *Le Livres des Etoiles* is a Latin translation of the Arabic *Book of Fixed Stars* by Al-Sufi, a thorough depiction of stars and constellations published in 964. The Arabic book was an attempt to achieve a synthesis of classical astronomy with indigenous Arabic tradition.

the astrolabe, a Greek invention for solving the problems of spherical astronomy. Islamic scholars like Al-Farghani wrote impressive treatises on how astrolabes could be applied mathematically to solve astronomical problems.

Ptolemy had identified over a thousand stars in his *Almagest*. This list was revised in 964 by Al-Sufi (*c.* 903–986), who carried out observations on the stars and described their positions, magnitudes, brightness, and color in his *Book on the Constellations of Fixed Stars*. He provided two drawings for each constellation, one from the outside of a celestial globe and one from the inside.

Although the Islamic astronomers worked with the geocentric framework, they expanded on Ptolemy's studies. Muhammad al-Battani (*c.* 858–929) improved Ptolemy's astronomical calculations by replacing geometrical methods with trigonometry. He established the position of the solar orbit (equivalent to the earth's elliptical orbit now) more accurately than Ptolemy. He predicted eclipses, worked on the phenomenon

of parallax, and calculated the values for the precession of the equinoxes and the inclination of the earth's axis, as also timings for the new moon and lengths for the solar year and sidereal year. His *Al-Zij al-Mumtahan*, a compendium of astronomical tables, is considered the most important astronomical work between Ptolemy and Copernicus.

Ibn Yunus (*c.* 950–1009) made very accurate astronomical observations in his *Zij al-Kabir al-Hakimi*. He described 40 planetary conjunctions and 30 lunar eclipses and observed more than 10,000 entries for the sun's position for many years

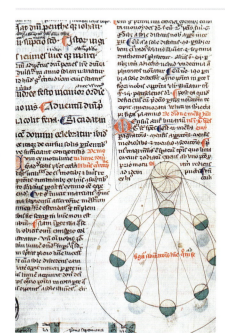

**MOON PHASES,** *De Arte Matematicum*, 15th century

This page from a medieval manuscript explains the phases of the moon. The sun (at center right) is shown illuminating one side of the moon as it moves in its orbit. Both the sun and moon are shown moving in circular orbits around the earth. This diagram shows that the moon will appear full when the moon and sun are on opposite sides of the earth, while it will appear darkened when they are on the same side of the earth.

# ZIJ

Zij is the generic name for Arabic astronomical tables used for calculations of the positions of the sun, the moon, stars, and planets. Over 200 zijes are known to have been produced by Arabic astronomers between the 8th and 15th centuries.

The first tables discovered by the West were from Al-Khwarizmi. They were translated into Latin by Adelard of Bath in 1126 and subsequently modified for the Christian Era and for various European meridians. Later they were joined by the Toledo Tables, credited to the 11th-century astronomer Al-Zarqali or Arzachel (1028-1087), whose translated canons were particularly popular. In the 1260s, Alfonso X of Spain (1221-1284) ordered the compilation of tables designed to be universal. These tables, called Alfonsine Tables, allowed the user to adjust for meridian and epoch. Extant only in the form given them by Johannes de Lineriis and his student Johannes de Saxonia in Paris in the 1320s, the Alfonsine Tables remained the standard for European astronomy until the 16th century.

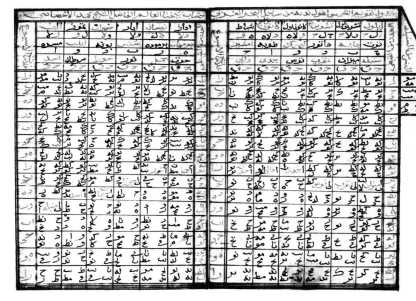

**HISTORICAL TABLE OF ASTRONOMICAL OBSERVATIONS OF THE SUN,** From *The Shining Pearl* or the *Operations of the Sun* by Muhammad al-Akhasai al-Muwaqqit, 8th-17th century

Observatories from Damascus to Baghdad, and later Spain, took detailed observations of the sky, including the sun.

using a large astrolabe. In 1030, Abu al-Rayhan al-Biruni (973-1048) said that the earth rotated on its own axis.

Although the Islamic astronomers accepted the Ptolemaic-Aristotelian cosmology of tightly-nested spheres centered on the earth, doubts began to emerge in the 10th century. Ibn al-Haytham was one of the first critics who, in his *Doubts on Ptolemy*, said that Ptolemy's equant did not satisfy the requirement of the uniform circular motion and declared Ptolemy's planetary models false. Abu Said al-Sijzi (c. 945-1020), a contemporary of Al-Biruni, suggested the possible heliocentric motion of the earth around the sun. In 1070, Abu

Ubayd al-Juzjani (11th century), addressing the equant problem in Ptolemy's model, proposed a non-Ptolemaic configuration in his *Tarik al-Aflak*.

In the 12th century, Ibn Rushd (1126-1198) of Andalusia, known by his Latin name Averroës, rejected Ptolemy's eccentric deferents and argued for a strictly concentric model of the universe. This started the Andalusian Revolt, as the astronomers in Andalus set about developing a non-Ptolemaic model. Ibn Bajjah or Avempace (c. 1095-1138) was the first to discover a system in which no epicycles occur. Later in the 12th century, his successors Ibn Tufail (c. 1105-1185) and Nur Ed-Din

**IBN RUSHD** or Averroës, Córdoba, Spain

Ibn Rushd started a revolution in astronomy when he rejected the Ptolemaic model and argued for a strictly concentric model of the universe. He wrote a commentary on Aristotelian physics and was the first to define and measure force.

Al Betrugi (*d. c.* 1204) proposed planetary models without any equant, epicycles, or eccentrics.

In the 13th century, Nasir al-Din al-Tusi (1201-1274), finding the equant particularly faulty, replaced it in his *Tadhkira* (*Memorandum*) by adding two more small epicycles to each planet's orbit. He gave a geometric construction, the Tusi couple, to modify the Ptolemaic planetary model and founded an observatory at Maragah. This was the beginning of the Maragah school of astronomers who were more successful than their Andalusian predecessors in producing non-Ptolemaic configurations, which eliminated the equant and eccentrics. Some of the prominent Maragha astronomers included Mo'ayyeduddin Urdi, 'Umar al-Katibi al-Qazwini, Qutb al-Din al-Shirazi, Sadr al-Sharia al-Bukhari, Ali al-Qushji, al-Birjandi, and Shams al-Din al-Khafri. Finally, in about 1350, Ibn al-Shatir (1304-1375) achieved a completely concentric rearrangement of the planetary mechanisms, eliminating the equant and other objectionable circles of Ptolemy's model. It is believed that the Maragha models provided inspiration to Copernicus, who mentions the theories of Al-Battani, Arzachel, and Averroës as influences in his *De revolutionibus orbium coelestium* (*On the Revolutions of the Heavenly Spheres*) in the 16th century.

## DYNAMICS IN MEDIEVAL EUROPE

From the 12th century onward, the translations of the Arabic writings familiarized the Latin West with the Ptolemaic system and the physical and metaphysical doctrines of Aristotle. In the 13th century, theologians at the University of Paris were disturbed by the statements in Aristotle's theory that implied limitations of God's powers. In 1277, Étienne Tempier (*d.* 1279), the Bishop of Paris, acting on the advice of the theologians of Sorbonne, criticized a great number of errors in the philosophy of the Peripatetics, the school of Aristotle.

Among the "errors" considered dangerous to faith was the opinion maintaining that God himself cannot give the entire universe a rectilinear motion as the universe would then have a vacuum behind it. Aristotle's physics treated the existence of an empty space as a pure absurdity. Around 1280, Richard of Middletown (*c.* 1249–1302), and after him many other masters at Paris and Oxford, admitted that the laws of nature are certainly opposed to the production of empty space. A branch of physics known as dynamics had to be created to explain motion in vacuum.

Before this, problems involving motion were not well understood, perhaps because of Aristotle's erroneous theory that motion required the continuous application of force, such as the air surrounding a moving object. Jean Buridan (c. 1295-1358), who taught at the University of Paris, corrected Aristotle's theory of motion by developing a theory of impetus, the first step toward the concept of inertia. According to this theory of motion, the mover imparts to the moved a power or impetus, proportional to the velocity and mass, which keeps it moving. In addition, he correctly established that the resistance of the air or other medium progressively reduces the impetus and that weight can add or detract from speed. Buridan explained the motion of the heavens through this concept.

In the 1330s, scholars at Oxford studied motion in constant acceleration and concluded that in a uniformly accelerated motion, distance increases as the square of time. This came to be known as the Merton acceleration theorem. Buridan's student Nicole Oresme (c. 1323-1382) gave a geometric proof of this around 1361. He wrote a commentary on Aristotle's treatise *De Caeto* in 1377, maintaining that neither experiment nor argument could determine whether daily motion belonged to the firmament of the fixed stars or to the earth. Oresme also offered several considerations favorable to the hypothesis of the earth's daily motion. This also led him further to explain how in spite of this motion, heavy bodies fell in a vertical line. He suggested that the speed of fall is directly proportional to the time (and not the distance) of fall. This is exactly the principle that Galileo was to discover later.

The Great Western Schism (lasting from 1378 to 1417) halted the progress of science and led to the diffusion of Parisian learning as the masters at the University of Paris fled to other countries. In 1386, Marsilius of Inghen (c. 1330-1396), who was one of the most gifted professors of the University of Paris, became Rector of the nascent University of Heidelberg, where he introduced the dynamic theories of Buridan. Around the same time, another master of Paris, Henry of Langenstein, also known as Henry of Hesse the Elder (c. 1325-1397), was instrumental in founding the University of Vienna and brought to it the astronomical tradition

**A LESSON IN THEOLOGY AT THE SORBONNE**, From *Postilles sur le Pentateuch*, Bibliothèque Municipale, Troyes

The Sorbonne, founded by Robert de Sorbon about 1257, grew out of the cathedral schools of Notre-Dame and soon became the most celebrated teaching center of Christian orthodox theological teaching. Its famous masters included Alexander of Hales, St Bonaventure, Albertus Magnus, and Thomas Aquinas. The theologians of Sorbonne condemned many Peripatetic doctrines as blasphemous, giving rise to new theories in physics. Jean Buridan revised Aristotle's theory of motion by presenting the theory of impetus, arguing that a projectile continues in motion not, as Aristotle held, because it is supported by the surrounding air, but because of the force transmitted to it by the object that launched it. The impetus school flourished in Paris during the 14th and 15th centuries and included Albert of Saxony, Nicholas Oresme, and Nicholas of Cusa.

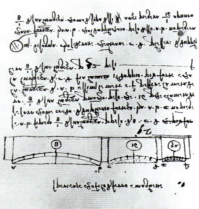

## LEONARDO DA VINCI'S NOTES

**Far Left:** Notes by Leonardo da Vinci on the flow of water. The notes are in mirror writing, a technique Vinci used to keep his ideas secret. The diagrams show swirling geometrical patterns formed by water flowing past a stationary object. His notes compare these patterns with those seen in braided hair.

**Left:** Leonardo da Vinci's drawing of the effect of forces on beams supported at both ends (1490). Vinci dealt extensively with mechanics, writing on tensile strength, analysing forces produced by arches and hoisting devices, and examining the strength of beams and columns.

established by Ptolemy. By perfecting all the details of Ptolemy's theories, Henry's successors were helpful in bringing to light the defects of the theories and thereby preparing the materials by means of which Copernicus was to build the "new astronomy."

## BEGINNING OF THE RENAISSANCE

The Hundred Years' War (1337-1453), the Black Death, and other disasters halted the progress of science in the 14th century. But the rediscovery of ancient scientific texts after the Fall of Constantinople in 1453, and the invention of printing—which allowed a faster propagation of new ideas—stimulated a new interest in science. Yet humanists favored human-centered subjects like politics and history over natural philosophy or applied mathematics. As a result, the development of science was comparatively slow at the beginning of the Renaissance.

At the beginning of the 15th century, masters such as Paolo Nicoletti propagated the concept of dynamics in the Italian Universities of Padua and Bologna. Nicholas of Cusa (1401-1464) was greatly influenced by these studies and gave theories of the universe that anticipated the works of Kepler and Copernicus.

Leonardo da Vinci (1452-1519) was perhaps more convinced about the merits of the Parisian physics than any other Italian master. A versatile genius with insatiable curiosity, he had studied a great number of works by authors such as Albert of Saxony, Nicholas Oresme, and Nicholas of Cusa. He initiated a new idea—the composition of concurrent forces—which may be simply stated as follows: the two component forces have equal moments as regards the direction of the resultant. Also, the resultant and one of the components have equal moments as regards the direction of the other component. Based on his study of Albert of Saxony's theory of the center of gravity, Vinci could determine the centre of gravity of a tetrahedron. He also presented the law of the equilibrium of two liquids of different density in communicating tubes. He was able to discover the same laws of hydrostatics that the French scientist Pascal did. The elastic reactions of deformed bodies and the laws governing them, formulated by Buridan, Albert of Saxony, and Marsilius of Inghen, were applied by him in such a way as to draw from them an explanation of the flight of birds.

Sparks like Nicholas of Cusa and Vinci caused a stir in science, leading to the scientific revolution of the 16th century.

# Chapter 3
# SCIENTIFIC REVOLUTION

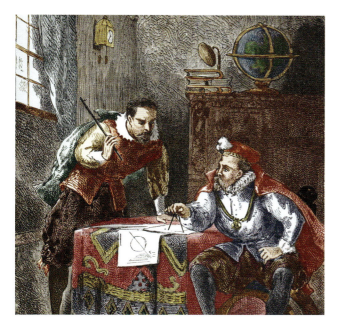

During the 15th, 16th, and 17th centuries, scientific thought was revolutionized. The medieval scientific philosophy was abandoned in favor of the new methods proposed by Bacon, Galileo, Descartes, and Newton. The development of the experimental method to seek answers to the problems of nature and the pursuit of science itself, rather than philosophy, marked the scientific revolution.

## ASTRONOMY

The scientific revolution started in astronomy with Nicolaus Copernicus (1473-1543), who overthrew traditional astronomy by proposing the heliocentric model of the cosmos. According to the Copernican theory, the sun, not the earth, is at the center of the universe; the earth, besides orbiting the sun annually, also turns once daily on its own axis; and that very slow, long-term changes in the direction of this axis account for the precession of the equinoxes. Though Copernicus put down his thoughts around 1510 in a manuscript called *Commentariolus* (*Little Commentary*), his final theory appeared in *De revolutionibus orbium coelestium libri vi* (*Six Books Concerning the Revolutions of the Heavenly Spheres*), printed only in 1543, the year of his death.

The implications of the Copernican revolution took some time to sink in. It was welcomed by professional astronomers and solved many problems of the Ptolemaic system. The central cosmological hypothesis of Copernicus faced widespread opposition in the church and outside, but astronomers, impressed by the observations and mathematical techniques employed by Copernicus, used *De revolutionibus* widely for solving advanced problems of astronomy. Yet it didn't answer all questions of the Ptolemaic system. For instance, it lacked an accurate description of the orbits of the planets. This necessitated the creation of a new science of celestial mechanics, which was established by Tycho Brahe (1546-1601), a Danish nobleman, and his assistant Johannes Kepler (1571-1630), a German astronomer.

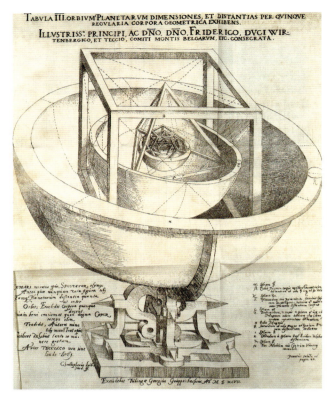

observations on the position of stars and planets that rendered all previous observations obsolete. Though unwittingly, his observations contradicted the Aristotelian system and reinforced the Copernican system.

Brahe's work was further explored and explained by Kepler, whose discoveries proved invaluable for the progress of science. Kepler's primary motivation behind his work *Mysterium Cosmographium* was a mystical desire to unravel the secrets of the universe. While Copernicus believed in cycles and epicycles like Ptolemy, Kepler discovered from his observation of the planet Mars that its orbit was an ellipse with the sun as its focus. He rejected the Pythagorean-Platonic ideal of uniform circular motion of the heavens and established the hypothesis of elliptical orbits and two other laws by which he explained the speed of planets in their orbits around the sun. His scientific approach ended the religious and philosophical prejudices that had been plaguing the world since the Middle Ages.

Brahe was able to use his influence with King Frederick II to build in 1576 the first scientific institute of the modern world, Uraniborg, on the island of Ven. There, with larger and more accurate apparatus, Brahe collected a series of exact

**1350-1600** Renaissance in Europe.

**1500-1600** The Protestant Reformation.

**1543** Copernicus publishes *De revolutionibus orbium coelestium*.

**1576** Tycho Brahe constructs a planetary observatory.

**1600-1800** The Age of Enlightenment, an intellectual movement in Europe.

**1609** Johannes Kepler presents his first and second laws of planetary motion.

**1610** Galileo Galilei observes the phases of Venus, moons of Jupiter, craters on the moon, and stars in the Milky Way with his telescope.

**1618-1648** Thirty Years' War in Europe.

**1619** Kepler presents his third law of planetary motion.

**1619** René Descartes' theory of rationalism.

**1620** Francis Bacon promotes the empirical scientific method.

**1621** Snell's law of refraction.

**1637** René Descartes presents his mechanical philosophy.

**1644** Evangelista Torricelli invents the mercury barometer.

**1648** Blaise Pascal explains barometer as a result of atmospheric pressure.

**1687** Isaac Newton publishes laws of motion.

**1690** Christiaan Huygens presents the wave theory of light.

**1704** Newton publishes *Opticks*.

## GALILEO'S ASTRONOMICAL OBSERVATIONS

**Left:** Galileo demonstrating his telescope, the first to be used for astronomical observations, to the Doge (ruler) of Venice, Italy, in August 1609. Galileo (on the right of the telescope) heard of the first telescope or "magic tube" invented by Lippershey in Holland in 1608 and built his own design in 1609.

**Right:** Sketches of the moon by Galileo, From *Sidereus Nuncius*, 1610. Using his telescope, Galileo saw that the terminator (the line between the moon's night and day sides) was sometimes irregular (top) and sometimes smooth (bottom). He deduced that the irregularities were due to mountains on the moon, the first time earth-like objects had been discovered in the heavens. This challenged the Aristotelian worldview that the heavens were perfect and unchanging.

Kepler's laws raised the fundamental physical question of what holds the planets in their orbits. He cited a force analogous to magnetism—introduced by William Gilbert in *On the Magnet, Magnetic bodies, and the Great Magnet of the Earth* in 1600—as the answer for this problem. Kepler stated that the sun emanated a magnetic force that pushed the planets in their orbits, though he couldn't quantify this theory. Nevertheless, his observations brought astronomy into the realm of physics and provided Isaac Newton in England, and Leibniz, Laplace, and Lagrange in France enough ground to arrive at quantitative and dynamic explanations. The scientific revolution had truly arrived.

Around the same time as Kepler, Galileo Galilei (1564-1642) of Padua, Italy constructed his own telescope to observe the heavens. His observations dismantled the Aristotelian system completely. He observed that the surface of the moon is mountainous, and not smooth as Aristotle had stated; that the earth doesn't have light of its own; that there are satellites circling Jupiter, so the earth is not unique in having its own satellites; and that planets orbit the sun and not the earth, as proved by the phases of Venus.

## MECHANICS

As an extension of his defense of the Copernican system, Galileo tried to arrive at the universal laws of motion through both mechanical and mathematical experiments, in the process establishing a new physics that was highly mathematized. He believed that motion—whether natural or artificial—had universally consistent characteristics that could be described mathematically. A story often told about Galileo is that he dropped bodies of different weights from the Leaning Tower of Pisa in order to demonstrate that all bodies under the influence of gravity fall with equal speed. This was contrary to Aristotle's theory that heavy objects fall faster than lighter ones, in direct proportion to weight. Whether the story is true or false, the fact remains that Galileo made the greatest scientific contribution in connection with the fall and motion of bodies. He proposed that a falling body would fall with a uniform acceleration, provided that it is falling through a vacuum or there is no resistance. He also derived the correct kinematical law of free fall—that the distance traveled by a falling body is proportional to the square of the elapsed time—and the curved (or parabolic) path followed

*ro esser eretto all' Orizonte, & il Prisma, ò Cilindro fitto nel muro ad angoli retti) è manifesto che douendosi spezzare si romperà nel luogo B, doue il taglio del muro serue per sostegno, e la B C per la parte della Leua, doue si pone la forza, e la grossezza del solido B A è l'altra parte della Leua, nella quale è posta la resistenza, che consiste nello staccamento, che s' hà da fare della parte del solido B D, che è*

by a missile or projectile. He also gave a principle of inertia, stating that objects retain their velocity unless impeded by friction, refuting the generally accepted Aristotelian hypothesis that objects naturally come to a halt once the initial force stops acting upon them. Thus Galileo established a new natural philosophical tradition—based on mechanical experimentation and mathematical description rather than a verbal, qualitative account—which found followers in Evangelista Torricelli and members of the Accademia del Cimento in Italy, Marin Mersenne and Blaise Pascal in France, Christiaan Huygens in the Netherlands, and Robert Hooke and Robert Boyle in England. His work on the motions of bodies was a precursor of the classical mechanics developed by Isaac Newton.

Evangelista Torricelli (1608-1647) assisted Galileo during the latter part of his life. It is probable that Galileo urged Torricelli to use mercury to investigate the so-called "horror of the vacuum." Torricelli filled a 4-ft (1.2-m) long glass tube with mercury and inverted the tube into a dish. He observed that the column of mercury only descended partially and deduced that

the space created by the descent of the mercury in the tube was vacuum. He also established that the pressure exerted by the air in the tube held up the column of mercury, concluding that air had weight. Based on this experiment, Torricelli built the first mercury barometer around 1644.

By 1646, Blaise Pascal (1623-1662) had learned of Torricelli's experiment with barometers. At this time, most scientists believed that, rather than a vacuum, some invisible matter was present above the column of mercury in the tube. This was based on the Aristotelian notion that vacuum was an impossibility. Pascal replicated Torricelli's mercury experiment and proved the hypothesis that it was vacuum above the column of mercury in the tube and that air had weight. Following more experimentation in this vein, in 1647 Pascal published *Experiences nouvelles touchant le vide* (*New Experiments with the Vacuum*), an account of his experiments to determine what degree of mercury and water could be supported by air pressure. It also provided reasons why it was indeed a vacuum above the column of liquid in a barometer tube. His brother-in-law Florin Périer helped him to measure air pressure with a barometer as he climbed Puy-de-Dôme, a 4,806-ft (1,464-m) high mountain in France. In his *Treatise on the Equilibrium of Liquids and Treatise on the Weight of the Mass of the Air*, Pascal concluded, "Since the weight of the air produces all the effects hitherto attributed to the horror of the vacuum, it should be the case that since this weight is not infinite but has limits, its effects too should be limited, which is confirmed by experiment."

Around this time, René Descartes (1596-1650) proposed the theory of matter and motion, known as the mechanical philosophy, which filled the void left by the newly-displaced Aristotelianism. According to Descartes' vision of the universe, God sets matter in motion and imposes the laws of nature upon it. As per the principle of inertia, matter, once set in motion, cannot stop, but must continue to move in a straight line until something else stops it. In the Cartesian universe, matter moves as a vortex, which are large circling bands of minute particles, and these particles of matter collide continuously with one another. All natural phenomena are the result of these collisions, so it is important to study the quantitative laws of impact. This was achieved by Descartes' successor, Christiaan Huygens (1629-1695), who formulated the laws of conservation of momentum and of kinetic energy.

Isaac Newton (1642-1727) reacted to Descartes' science with his monumental *Philosophiae Naturalis Principia Mathematica* (*Mathematical Principles of Natural Philosophy*), published in 1687. Originating as a treatise on the dynamics of particles, the *Principia* presented an inertial physics that combined Galileo's mechanics and Kepler's planetary astronomy. Newton invented

**EVANGELISTA TORRICELLI WORKING ON THE BAROMETER,** From *The Atmosphere* by Camille Flammarion, 1873

Torricelli is best remembered for his invention of the mercury barometer and his discovery of atmospheric pressure in 1644.

### PASCAL'S EXPEDITION

Here Pascal's brother-in-law Florin Périer and two friends are seen carrying out an experiment in air pressure on the physicist's behalf. Pascal wanted to prove Torricelli's hypothesis that air pressure would decrease with altitude. To do this, he requested his brother-in-law to carry a mercury barometer to the summit of the Puy-de-Dôme mountain, France. The barometer recorded a drop in air pressure as the party made their ascent, thus proving Torricelli's hypothesis.

calculus in mathematics, which enabled him to cast the laws of motion, especially acceleration and gravity, in an analytical form. He gave mathematical solutions to the problems in mechanics and astronomy, and thus synthesized celestial and terrestrial physics. In the *Principia* Newton wrote with pride, "I now demonstrate the frame of the system of the world." He mathematically derived all three of Kepler's laws of planetary motion, which were necessary to understand the laws of gravitation. He was able to discover these laws by observing the fall of an apple on the ground, which he equated with the fall of the moon toward the earth. Kepler and Newton represented a critical transition in human history with the fundamental conclusions that the rules that apply on earth also apply to the universe. That is why their laws of gravitation are considered to be universal.

Unlike Aristotle who philosophized physics, Newton used experimental and mathematical methods to explain natural phenomena. Though the *Principia* may be described as a work of experimental physics, mathematical physics, or the

science of nature, Newton preferred to use the term "natural philosophy." He laid emphasis on "force" and was concerned with investigating "the forces of nature" in his three laws of motion. The first law relates to inertia and the tendency of a body to continue in its state of rest or uniform motion unless it is disturbed by an external force.

Galileo had observed, ". . . any velocity imparted to a moving body will be rigidly maintained as long as the external cause of acceleration or retardation are removed, a condition which is found only on horizontal planes; for, in the case of planes that slope downwards there is already present a cause of acceleration, while on planes sloping upwards there is retardation. From this it follows that motion along a horizontal plane is perpetual, for, if the velocity is uniform it cannot be diminished or slackened much less destroyed." Newton saw things in a larger context than Galileo. He observed that uniform straight line motion is inertial and that such motion is not confined to a horizontal plane. Newton recognized that it takes force to disturb a body from its inertia, which he

**BERNOULLI BROTHERS**

Swiss mathematician brothers Johann and Jakob Bernoulli are seen debating a mathematical problem in 1696. Johannn (right), who wrote on differential equations and curves, posted the famous brachistochrone problem to a group of mathematicians. In essence, the challenge was to find the curve along which a particle will most quickly move from one point to another which is not directly below it. Jakob (left), who studied infinite series, curves, and differential calculus, devised a correct proof, leading to a cycloid curve and the calculus of variations. Applications of variational principles occur in theories of motion, quantum mechanics, elasticity, electromagnetic theory, aerodynamics, the theory of vibrations, and other areas in engineering and science.

**STAMPS COMMEMORATING 300 YEARS OF *THE PRINCIPIA*, 1987**

Newton published his revolutionary *The Philosophiae Naturalis Principia Mathematica* in three volumes in 1687. The work contains Newton's three laws of motion, forming the foundation of classical mechanics, as well as his law of universal gravitation and a mathematical derivation of Kepler's laws for the motion of the planets.

quantified as mass times acceleration. Acceleration is the rate of change of velocity in every direction. He applied the tools of calculus to specify this law, known as the second law of motion. Leonhard Euler of Switzerland (1707-1783) wrote the second law in its current form: $F = ma$, where $a$ is acceleration. So the change in velocity is directly proportional to the force acting on it and inversely proportional to its mass.

Newton's third law of motion states that "to every action there is always an equal and opposite reaction." While Galileo had anticipated the first two laws, the third law was unique to Newton. At a cursory glance it may appear as if Newton's third law would prevent any motion at all, but deeper reflection shows why this is not so. To cite an example, when a rocket ignites its motors, exhaust gases from the propellant act on the earth, and the rocket moves up against gravity, in direct relationship to the gases being expelled. Gravity is nothing but acceleration that acts toward the center of the earth. Newton's contributions to the universal law of gravitation that governs all motion in the universe was an extension of the experimental facts established by Kepler and Copernicus.

In 1696, Swiss mathematician and hydrodynamic expert Johann Bernoulli (1667-1748) challenged his colleagues to solve the famous brachistochrone problem. The problem was solved independently by Newton, Bernoulli and his brother Jakob (1654-1705), and led to the invention of the calculus of variations, which was later applied to the measurement of curves, to differential equations, and to mechanical problems. The calculus of variations led Joseph-Louis Lagrange (1736-1813) in France to discover analytical mechanics.

## OPTICS

Kepler was the first to explore optics in the 17th century with his *Paralipomena* in 1604. His exploration of the astronomical phenomena—such as how a ray of light, coming from a heavenly body located in the outer regions of space, deflects when entering the denser atmosphere around the earth—led him to study the behavior of light in immediate surroundings—that is, what happens to light as it enters the relatively denser medium of the human eye. He was the first to explain the process of vision by refraction within the eye, the first to give a geometric theory of lenses, the first to use a pin-hole camera to investigate the formation of pictures, the first to formulate eyeglasses for nearsightedness and farsightedness, and the first to explain the use of both eyes for depth perception. He also gave the first mathematical account of Galileo's telescope.

In 1621, Willebrord Snellius (1580-1626) discovered the Snell's law of refraction, which was mathematically derived by Descartes in 1637 in *Discourse on Method*. However, it was Newton who made the most important contribution to optics in the 17th century. Newton found that he could derive Snell's law of refraction if he made these assumptions: one, light consists of very small bodies called corpuscles; two, refraction is caused by a force of attraction exerted on these light particles in a direction perpendicular to the plane between the two media and depending in amount only on the distance from that separating plane. However, the core of Newton's contribution was his theory of colors. Traditional theories considered colors to be modification of light, believed to be white in its pristine form. Through a series of experiments performed in 1665 and 1666, in which the spectrum of a narrow beam was projected onto the wall of a darkened chamber, Newton rejected the traditional theory by demonstrating that white light is a mixture from which different beams of color could be separated. He also concluded that rays of different colors refract at different degrees; that's how a rainbow is formed or prisms produce spectra of colors from white light. His work on the reflecting telescope and discovery of the inference phenomena, called Newton's rings, are his other important contributions to optics. Newton believed that light always traveled in a straight line without bending around an obstacle. This characteristic of light is consistent with the corpuscular nature of light. He published his experiments on the nature of light in *Opticks* in 1704.

The wave theory of light developed by Christiaan Huygens (1629-1695) of Holland around the same time as Newton challenged the corpuscular theory. Huygens published his *Treatise on Light* in 1690, arguing that it was not possible for solid corpuscles to travel at the speed of light. It would not be wrong to say that Huygens was a visionary who set in motion the theory that could explain the phenomena of light, sound, and heat in mathematical terms. Like sound needs a medium to propagate it, Huygens assumed that light too needs a medium—luminiferous ether—to propagate.

In the absence of enough experimental evidence, Newton's corpuscular theory of light was favored over the wave theory

**NEWTON'S OPTICS**

Newton conducting his famous experiment on light, using a prism to refract a ray of light from a hole in the shutters over a window.

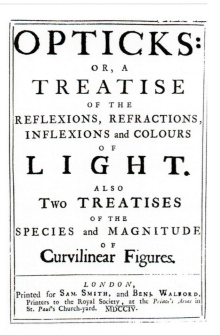

Title page of Newton's *Opticks*, published in 1704

Fresnel developed his own wave theory of light and presented it to Académie des Sciences in 1817. He performed new experiments on interference and polarization effects, and showed that to explain the effects light must be in the form of transverse waves as Young and Huygens had suggested, rather than longitudinal waves. Fresnel applied his skills to the development of lenses for light-houses.

**CHRISTIAAN HUYGENS PRESENTING A CLOCK TO LOUIS XIV**

The idea of using a pendulum to measure time had been proposed by Galileo but he had not found how to keep one swinging. In 1656, Huygens designed a table clock that kept the pendulum swinging by means of a spring. Huygens moved to France to work for Louis XIV during 1666-1681 and published the results of his clock research in *Horologium Oscillatorium* (1673). He founded the wave theory of light, contradicting Newton's particle theory of light, and demonstrated how waves might interfere to form a wavefront, propagating in a straight line. He opened the field of physical optics dealing with the phenomena of interference and diffraction of light.

for more than a century, till Thomas Young (1773-1829) and Augustin-Jean Fresnel (1788-1827) conducted experiments on interference and diffraction of light. They showed that light waves could interfere with each other like sound waves and bend around an obstacle, and that light could be polarized—a phenomena that could only be explained by the wave theory. While Newtonian theory implied that light travels more swiftly in a denser medium, the wave theory of Huygens stated the opposite. In 1850, Léon Foucault of France (1819-1868) measured the speed of light in air and water and proved the wave theory to be right.

The wave theory received additional support from the electromagnetic theory of Michael Faraday and James Clerk

Maxwell in the 19th century, who showed that light is a form of electromagnetic radiation, a theory confirmed by Heinrich Hertz (1857-1894) in 1886 with the discovery of radio waves. The electromagnetic theory of light was a great triumph for 19th-century physics. However, it still left some questions unanswered, which were resolved by Albert Einstein in 1905. His special theory of relativity answered the questions related to the speed of light. To explain the photoelectric effect, he expanded the quantum theory of thermal radiation proposed by Max Planck in 1900. He said that light travels as tiny bundles of energy called light quanta, or photons, and thus resurrected Newton's particle theory of light. However, he did not reject the wave theory of light.

The theories of Newton and Huygens collectively shaped the modern theory of wave-particle duality, based on the hypothesis that light has both wave- and particle-like attributes. In 1924, Louis de Broglie (1892-1987) proved the wave-particle duality experimentally. The quantum mechanical theory of light and electromagnetic radiation continued to evolve through the 1920s and 1930s, and culminated in the development of quantum electrodynamics (QED) in the 1940s.

# Chapter 4
# MODERN PHYSICS

**MICHAEL FARADAY,** 1840

British scientist Michael Faraday, who contributed significantly to the study of electromagnetism and electrochemistry, is demonstrating an electrical apparatus to workmates at a bookshop.

**FACING PAGE:** Albert Einstein explaining the mass-energy equation, which resulted from his 1905 paper on Special Relativity and formed the basis of nuclear fusion and atomic bombs.

Newton contributed to the climate of Enlightenment in Europe by giving rational and empirical explanation for all natural phenomena. The *Principia* established a mathematical approach to physics, leading to the development of celestial mechanics in the 18th and 19th centuries. The leading exponents of this branch of astronomy included Alexis Clairaut, the Bernoulli family, Leonhard Euler, Jean Le Rond d'Alembert, Joseph-Louis Lagrange, and Pierre-Simon Laplace, who developed and applied the calculus of variations for solving the astronomical problems in Newtonian mechanics. The *Opticks* served as a model of experimental investigation, influencing the studies on heat, light, electricity, and magnetism.

After the great religious and political disturbances of the previous 100 years, the latter half of the 17th century was a period of relative calm and active prosperity. It was an age of building of civilization. Scientists were recognized and honored. The governments and the ruling classes of all the leading countries had certain common interests in trade and navigation. These interests propagated culminating achievements of the third phase of the scientific revolution, in which science was used in an organized and conscious way for practical ends. The Industrial Revolution was certainly not a product of scientific advance in its first phase, though certain contributions in science, such as the steam engine, were the essential ingredients in its success. It was no accident that the intellectual formulations of science, the technical changes in industry, and the economic and political domination of capitalism grew in this period.

The 18th and 19th centuries were the greatest formative centuries of the modern world. These were the centuries that can be characterized as the liberating phase of human development in which humans had found the ultimate path to prosperity and unlimited progress. In the 19th century, science became a major agent for effecting technical developments. However, it was only in the 20th century that science became fully integrated into the productive mechanism.

# Electricity and Magnetism

During the 18th century, everyone was fascinated by electricity. In the previous century, the works of William Gilbert (1540-1604), who coined the term electricity, and Otto van Guericke (1602-1686), who built the first electric generating machine, had popularized experiments with electricity. In 1729, Stephen Gray (1666-1736), an English scientist, discovered conductivity. In France, Charles François de Cisternay DuFay (1698-1739) performed many experiments based on Gray's work. In 1733, he announced that electricity consisted of two fluids: vitreous (positive) and resinous (negative). He also discovered that like charges repel and opposite charges attract. Dutch physicist Pieter van Musschenbroek (1692-1761) revolutionized the study of electrostatics in 1745 by his invention of the Leyden jar, a device that could store large amounts of charge. In the United States, Benjamin Franklin (1706-1790), known for his 1752 kite experiment, disagreed with DuFay's two-fluid theory and gave the single-fluid theory of electricity. Franklin encouraged Joseph Priestley (1733-1804) to publish *History and Present State of Electricity* in 1765, a compilation of all available data on electricity. From his experiments on electricity, Priestley anticipated the inverse square law of electrical attraction, which was confirmed by Charles Coulomb (1736-1806) in 1785 by a direct measure (using a torsion balance) of the force

**VOLTA SHOWING HIS VOLTAIC PILE TO NAPOLEAN** From *Le Petit Journal*, Paris, 1801

Volta constructed a device in 1800 to show that electric current was produced when two dissimilar metals were brought into contact. He demonstrated a compact version of this device—made of a stack of discs of copper, zinc, and cardboard moistened with salt solution—to Napolean in 1801.

**1700s** Enlightenment – origin of modern scientific disciplines.

**1752** Benjamin Franklin proposes the single-fluid theory of electricity.

**1800** Alessandro Volta invents chemical batteries and voltage.

**1820** Hans Christian Ørsted notices that electric current deflects a magnetized needle.

**1821** Michael Faraday shows that changing magnetic field produces electricity.

**1840** James Joule and Hermann von Helmholtz propose that electricity is a form of energy.

**1844** Ludwig Boltzmann's statistical mechanics and the meaning of entropy.

**1873** James Maxwell presents equations of electromagnetism.

**1887** Heinrich Hertz transmits radio waves.

**1895** Wilhelm Roentgen discovers X-rays.

**1897** J.J. Thomson discovers the electron.

**1898** Pierre and Marie Curie separate radioactive elements.

**1900** Max Planck invents quantum theory of radiation.

**1905** Albert Einstein explains Brownian motion and formulates photon theory of light and the theory of Special Relativity.

**1913** Niels Bohr applies Planck's theory to the atomic structure.

**1915** Einstein formulates General Theory of Relativity.

**1923** Erwin Schrödinger's wave equation.

**1927** Max Born and Werner Heisenberg formulate matrix mechanics.

**1949** Richard Feynman, Julian Schwinger, and Tomonaga Shin'ichirō formulate quantum electrodynamics.

**1967** Steven Weinberg, Sheldon Glashow and Abdus Salam's unified theory of electromagnetic and weak forces.

**1964** Superfluid helium-3 discovered by David Lee, Douglas Osheroff, and Robert Richardson.

**1986** Discovery of high-temperature superconductivity by Karl Müller and J. Bednorz

between two charged conductors. He transformed Priestley's descriptive observations into a quantitative law of electric force, known as the Coulomb's law, which states that the force between two electrical charges is proportional to the product of the charges and inversely proportional to the square of the distance between them. This law became the basis of the theory of magnetostatics developed by Siméon-Denis Poisson (1781-1840) of France.

The invention of the voltaic battery at the beginning of the 19th century by Alessandro Volta (1745-1827) provided a source of continuous electric current. Volta made an electrolyte cell by using two discs of copper and zinc, separated by a cardboard soaked in brine (salt water). He developed what is now called voltaic pile by connecting two or more cells in a stack. When he joined copper and zinc with a wire, electricity flowed continuously through the wire. Volta thus had a chemical source that produced continuous electric current. This was a rudimentary version of the battery.

The invention of battery allowed scientists like Georg Simon Ohm (1789-1854), a German, to study the flow of electricity quantitatively. He discovered the relation between voltage, current, and resistance in a circuit using direct current. The result, known as the Ohm's law, states that the current flow through a conductor is directly proportional to the voltage and inversely proportional to the resistance, or $I = V/R$.

Hans Christian Ørsted (1777-1851), a Danish physicist, established a relation between electricity and magnetism when he noticed that when a wire carrying electric current was placed near a magnetic needle, the needle swung at right angle to the wire. In 1820 he enthralled a whole generation of scientists with his discovery that electric current has magnetic effects, setting off a series of experiments in electromagnetism, a new field of physics. André-Marie Ampère (1775-1836), a French physicist, developed Ørsted's observations in quantitative terms by establishing the laws of magnetic force between electric currents. He also gave the right-hand rule for the direction of the force on a current in a magnetic field. Based on Ørsted's discovery, William Sturgeon of England and Joseph Henry of the US developed electromagnets in the 1820s. Now scientists began questioning if the inverse of the effect discovered by Ørsted was possible, that is, if magnetism could induce electric current.

A year later, in 1821, Michael Faraday built an electric motor that converted electricity into mechanical motion. Ten years later,

## JAMES MAXWELL

Maxwell is best known for his work on light and electromagnetic waves. He showed that oscillating charges produced waves in an electromagnetic field and that these waves had the same speed as light. He also predicted the existence of other forms of electromagnetic radiation, such as radio waves.

$$\nabla \times E = -\frac{1}{c}\frac{dB}{dt}$$

$$\nabla \times B = \frac{\mu}{c}\left(4\pi i + \frac{dD}{dt}\right)$$

$$\nabla \cdot D = 4\pi\rho$$

$$\nabla \cdot B = 0$$

## MAXWELL'S EQUATIONS

These equations, which relate electric and magnetic fields to current, were deduced by Maxwell in 1902. They predict that any change in an electric or magnetic force sends electromagnetic waves through space— a theory proved by the discovery of radio waves in 1888.

he designed the first electric generator (dynamo), in which a coil, made of copper, rotates in a magnetic field and produces current. Faraday's fascination led him to discover further linkage between electricity and magnetism. He discovered that iron filings, when placed near an electric current, formed the same pattern as that when placed near a magnet. He called the curved pattern created by the filings "a field" or line of force.

Scottish physicist James Clark Maxwell (1831-1879) wrote a letter to Faraday in 1857 in which he observed that "your line of force can weave a web across the sky and lead the stars in their courses without any necessarily immediate connection with the objects of their attraction." Newton had described gravitational force as "action at a distance" but was frustrated that he couldn't figure out how gravity works. Could Faraday's "line of force" provide the clue? Was there such a thing as "action at a distance"? While thinking about the line of force, Maxwell was to learn about some experiments by two German scientists, Wilhelm Eduard Weber and Rudolph Kohlrausch, who showed that the speed of electric current moving in a wire is close to the speed of light. Could therefore there be a connection between light, electricity, and magnetism? According to Newton's laws of motion, there is no limit as to how fast light or anything can go if enough force is applied to it. Maxwell questioned this instantaneous idea of Newton and wondered if light was an electromagnetic undulation or wave. He published a paper to set the record straight and concluded: light has a definite and set velocity in a vacuum. By using the experimental results, especially those of Faraday, Maxwell was able to establish rather convincingly the relationship between light and electromagnetism. He founded the four Maxwell equations that were to pave the way for Einstein's theory of relativity. This theory extended Newton's laws of motion and showed that Maxwell's equations for electromagnetism do not change under relativistic Lorentz's transformations. In other words, Newton's laws of motion do not apply to electromagnetism.

**JAMES JOULE MEASURING THE HEAT GENERATED IN WIRES FROM THE PASSAGE OF ELECTRIC CURRENT,** From *Physique Populaire*, 1891

Joule was interested in the physical properties of heat and, in 1840, determined that the heat generated in a wire by an electric current was proportional to the resistance and the square of the current (Joule's Law). In 1843, he determined the amount of mechanical work required to produce a given amount of heat. This work established the equivalence of heat and other forms of energy, and the principle of conservation of energy: energy is neither created nor destroyed.

Around this time, scientists were formulating relationships between electricity, magnetism, and other forms of energy. German scientist Hermann von Helmholtz (1821-1894), and the English physicists William Thomson (later Lord Kelvin; 1824-1907) and James Prescott Joule (1818-1889) investigated the relationship between electric currents and other forms of energy. In the 1840s, Joule established that the various forms of energy—mechanical, electrical, and heat—are the same and interchangeable, and thus formed the basis of the law of conservation of energy, the first law of thermodynamics.

# ELECTROMAGNETIC TECHNOLOGY

Maxwell's equations provided the theoretical basis for understanding the propagation of electromagnetic waves such as those which are received by antennas for use in radio and television. The world at large became fascinated with the new science of electromagnetism and the technology that was unraveled. When Samuel Morse (1701-1872) sent an electric current from Washington, D.C. to Baltimore in 1844, it turned a magnet on and off. When the magnet was on an iron lever, which pressed a rolling paper tape against an inked wheel, a long burst of current produced a dash; a short one made a dot. Morse used the technology to send a coded message—so the telegraph was born.

Thomas Alva Edison (1847-1931) changed night into day when he invented an incandescent light bulb in 1879. The tungsten filament lamp, introduced during the early 1900s, became the principal form of electric lamp, though fluorescent gas discharge lamps also became popular. Edison's employee Nikola Tesla developed the AC motor in 1888.

Alexander Graham Bell (1847-1922) invented the telephone in 1876. Working on Hertz's investigation of radio waves, the British physicist Ernest Rutherford (1871-1937) transmitted radio signals for more than half a mile in 1895. The Italian physicist Guglielmo Marconi (1874-1937) employed radio waves to send a wireless message across the Atlantic in 1901. Broadcast radio transmissions were established during the 1920s.

**NATIONWIDE US RADIO BROADCAST,** 1929

American Telephone and Telegraph Company radio engineers testing the installation for broadcasting the ceremonies for the inauguration of President Hoover in Washington, D.C., USA. This was only the second nationwide radio broadcast of a presidential inauguration.

The triode tube, invented by Lee De Forest of the US (1873-1961), made possible telephone transmissions by radio waves, the recording and reproduction of sound, and television. Known as the Audion, this device played a significant role in the development of the electronics industry.

Moving away from communications and broadcast entertainment, scientists and engineers in Britain, Germany, France, and the United States gave a new direction to electronics in the 1930s. They started research on radar systems capable of aircraft detection and antiaircraft fire-control. After World War II, the electronics industry peaked as the television became a commonplace, and a wide range of new devices and systems, especially the electronic digital computer, emerged.

The invention of the transistor and integrated circuit in the 20th century led to a diminution in the size and cost of electronic equipment, giving a further push to the industry. In the 21st century, electronic equipment has become far more sophisticated through the use of microchips, nanotechnology being the latest.

# Discovery of the Electron

The discovery of the electron in 1898 initiated a revolution in physics as big as Newton had created in the 17th century. In mid-19th century, Julius Plücker (1801-1868) improved the vacuum tube and sealed two electrodes inside the tube. When he passed electric current between the electrodes, a green glow appeared on the wall of the tube, which he attributed to rays emanating from the cathode. This started intensive research on the properties of cathode-ray discharges. In 1897, William Crookes of England (1832-1919) observed that the cathode rays cast a shadow and heated obstacles in the path. He was to conclude that they might be negatively charged particles, but no one was convinced.

In 1898, J.J Thomson (1856-1940), the English experimental physicist at the Cavendish Laboratory in Cambridge, UK, discovered that charge is carried by tiny particles named electrons. These tiny particles had a mass, which was 1/1840 of the hydrogen atom. Thomson observed that as the voltage

**J.J. THOMSON**

Expanding on the work with cathode rays, Thomson discovered the electron in 1898. He also measured their mass to charge ratio. For his proof of the existence of the electron, he was awarded the 1906 Nobel Prize for Physics. Later, he demonstrated the existence of isotopes.

between the plates is varied, electrons move toward the positive end of the plate. Going by the law that opposites attract while like repel, this established that electrons were negatively charged. When a horseshoe magnet is applied to the cathode ray tube, the negatively-charged beam deflects downward perpendicular to the magnetic field. The mechanical force that a charge experiences in motion perpendicular to the field is known as Lorentz force, after the Dutch physicist Hendrik A. Lorentz (1853-1928). According to

## STATIC ELECTRICITY

Electrons account for most of electric current, which can be described as a flow of electrons through a wire or conductor. Each electron carries one unit of negative charge, and each proton has the same amount of positive charge. When there is an excess of loose electrons, it means there is a negative charge and a deficit means there is a positive charge or excess of protons. Electrons flow from a region where there is an excess of them to a re-gion where there are fewer, that is, from negative potential to positive. When there is an imbalance between the negative and positive, electrons tend to jump the gap and cause a spark of static electricity.

When one combs hair or rubs feet on a carpet, one accumulates electrons and causes an imbalance of charge. If one touches a doorknob after that, there is a discharge of static electricity. Lightening, a big electric shock across the sky, is also caused by a discharge of electrons.

Here, a trickle of water is being bent by static electricity. The water is getting attracted toward the comb which is carrying an electric charge.

# X-RAYS

The German physicist Willhelm Roentgen (1845-1923) studied the "glow" or luminescence on the walls of Crookes' tube. When he covered the tube with black paper, the glow went right through the dark paper onto a screen in his laboratory. Then he found that when the cathode rays hit the element tungsten (used in light bulbs) it gave off a ray that passed through rubber, wood, and even his fingers! The year was 1895 and Roentgen had discovered X- rays that form the backbone of modern medicine.

Seventeen years later, Max von Laue (1879-1960), a German physicist, found that X-rays are electromagnetic waves like visible light but of much higher frequency. If Maxwell unified the electromagnetic waves equations, Max Planck in 1899 invented quantum mechanics that established a clear relationship between the energy of the radiation and the frequency of the wave (wavelength).

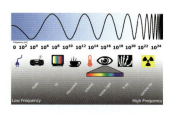

The electromagnetic spectrum (top); First-ever human X-ray, made by Roentgen in 1895.

Maxwell's equations, the electromagnetic waves move in a direction that is perpendicular to the electric field, which in turn is perpendicular to the magnetic field, hence the Lorentz force is a manifestation of the force the electron experiences from both these fields.

Thomson's discovery established the particulate nature of the electric charge and of the matter itself. He opened the door to the subatomic world and set the stage for Max Planck's quantum theory. Thomson's student Ernest Rutherford (1871-1937) of New Zealand eventually gave the model for the hydrogen atom.

## Radioactivity and the Atomic Structure

In 1896, the French physicist Antoine-Henri Becquerel (1852–1908) found strange unexpected rays emanating from a lump of uranium that he had left on top of a photographic plate wrapped in black paper inside a drawer. The mysterious rays marked the photographic plate while in the dark. These were not X-rays, which are electromagnetic. In 1895, Marie Curie (1867-1934) and her husband Pierre Curie (1859-1906) found that radium, like uranium, gave out similar rays. They called this phenomenon radioactivity.

Ernest Rutherford studied radioactivity and discovered two kinds of radiation: alpha rays (positively-charged particles

**PHYSICISTS WORKING WITH VACUUM TUBES,** From *Physique Populaire*, E. Desbeaux, 1891

The "radiant matter" physics was the cutting-edge physics of the day, with electric currents used to make gases glow in vacuum tubes. The Crookes tube (far left) developed from the Geissler tubes of the mid-19th century.

identical to the positive nucleus of helium) and beta rays (negatively-charged particles similar to the electrons discovered by Thomson). A third kind of radiation, gamma rays, is high-frequency electromagnetic radiation. Radium, uranium, and thorium, located toward the end of Mendelev's periodic table of the elements, are all radioactive and emit alpha, beta, and gamma rays from their interior. In 1902, while studying thorium, Rutherford and the English chemist Frederick Soddy (1877-1956) discovered that radioactivity caused changes inside the atom that transformed thorium into a different element. Based on their study, Rutherford and Soddy formulated the exponential decay law, which states that a fixed fraction of a radioactive substance will decay in a certain time period. For example, half of the thorium product decays in four days, half of the remaining sample in the next four days, and so on.

**THE CURIES**

Pierre Curie and wife Marie Curie discovered radium and polonium during their investigation of radioactivity.

The gold foil experiment of Hans Geiger and Ernest Marsden in 1909 led to the downfall of Thomson's plum-pudding model of the atom, in which negatively-charged "plums" are surrounded by positively-charged "pudding". Based on this experiment, Rutherford developed a nuclear model of the atom in 1911. In this model, atom has a dense, positively charged nucleus, in which all the mass is concentrated, around which negatively-charged electrons circulate.

## Thermodynamics and Quantum Theory

Thermodynamics and statistical mechanics involved the study of the relationship between heat, work, temperature, and energy, and their behavior in time. The first law of thermodynamics was related to the conservation of energy, as formulated in Galileo-Newton mechanics. Heat is thus a measure of the kinetic motion of molecules of a gas. Temperature is then nothing but a measure of the average energy of a molecule. Thus the field of statistical mechanics, unlike thermodynamics, operates on the mechanical motion at the micro level. The French engineer Sadi Carnot (1796-1836) enunciated the second law of thermodynamics at the microscopic level in 1815 to state that the steam engine cannot completely convert heat into work. The first law of thermodynamics states the opposite—conversion of work into heat is complete.

The German scientist Rudolf Clausius (1822-1888), a contemporary of Carnot, defined the concept of entropy to explain the second law of thermodynamics. He said that entropy creates unavailable energy and hence there cannot be 100 per cent efficiency to any thermodynamic engine. The Austrian physicist Ludwig Boltzmann (1844-1906) defined entropy as a measure of disorder. In a quantitative manner, Boltzmann defined entropy to be the logarithm of the thermodynamic probability. This probability was in turn described as the statistical state of a dynamical system.

**SADI CARNOT,**
Portrait by Boilly, 1813

In Carnot's time the efficiency of the steam engine was only about 5 per cent, meaning that 95 per cent of the heat energy of the burning fuel was wasted. Carnot gave the formula for deriving the maximum possible efficiency, which led to Clausius's formulation of the second law of thermodynamics and the concept of entropy.

**MAX PLANCK AND ALBERT EINSTEIN**

Working with Planck's quantum hypothesis, Einstein reached the conclusion in 1905 that light is composed of tiny bundles of light called quanta.

Lord Kelvin defined the absolute scale of temperature, that is, the common temperature measured in Celsius added with 273. At absolute zero, all disorder stops. This is the third law of thermodynamics, discovered by Walther Nernst (1864–1941) at Berlin at the beginning of the 20th century. Newton's classical mechanics and Huygens' construction of waves once again came up for consideration, except that temperature was now related to the average kinetic motion of molecules (or atoms). It also became the determinant of the characteristic of heat radiation.

In the 19th century, there were attempts to study radiation from hot bodies. Wilhelm Wien of Germany (1864–1928) had exhausted the resources of thermodynamics between 1890 and 1900 to arrive at a solution to this problem. Wien's law (maximum wavelength is inversely proportional to the absolute temperature of the body), while valid at high frequencies, was found to break down at low frequencies. Lord Rayleigh (John William Strutt; 1842–1919) and Sir James Hopwood Jeans (1877–1946) of Britain applied the science of statistical mechanics to deal with this problem. However, their results were not experimentally accurate.

Max Planck (1858–1947) of Germany combined thermodynamics and statistical mechanics into his study of blackbody radiation–electromagnetic radiation enclosed in a perfectly absorbing box (hence black) with walls that perfectly reflect the radiation back into the box. Planck referred to this box as *holraum* in German, one that was full of oscillations like a pendulum that absorbed and emitted radiation. Trying to derive Wien's law by the second law of thermodynamics, Planck arrived at the conclusion that the second law was not an absolute law of nature but a statistical law, as established by Boltzmann. Planck used thermodynamic theories of Joule, Kelvin, and Helmholtz as also the Maxwell-Boltzmann equipartition theorem, which appropriates equal amounts of energy to each different direction of motion of an atom. He reached the conclusion that oscillators comprising the blackbody could not absorb the radiation continuously but only in discrete amounts, in quanta of energy. In 1900, Planck developed the quantum theory of radiation, according to which radiation exists only in quanta of energy. Planck's theory, including the new universal constant h (his value was $6.55 \times 10^{-27}$ erg second, close to the modern value), laid the foundation of quantum mechanics and changed the theory of physical science.

# Einstein's 1905 Theories

Albert Einstein (1879-1955), a patent clerk in the Swiss post office, changed man's view of the universe in 1905, when he published four papers on the Special Theory of Relativity, the quantum theory of radiation, and a theory of the Brownian motion that led to the final acceptance of the atomic structure of matter. In two of the papers, Einstein presented a new method of counting and determining the size of the atoms or molecules in a given space and provided a theoretical explanation of the Brownian motion—the random movement of minute particles suspended in a liquid or gas. The idea that matter, such as a volume of gas, is composed of tiny particles, which are constantly in motion and collide with each other, was put forth in the kinetic theory of gases developed by Maxwell and Boltzmann in the 19th century. In an attempt to prove this theory, it was assumed that this motion would be imparted to larger particles that could be observed with a microscope. This line of reasoning led Einstein to combine the thermodynamics of liquids with statistical mechanics to obtain the first quantitative theory of Brownian motion. Jean-Baptiste Perrin's (1870-1942) microscope studies of Brownian particles confirmed Einstein's theory and established the atomic nature of matter.

The wave-theory of light did not explain the photoelectric effect—why dim blue light generates electric current when it strikes certain metals while bright red light fails to do so. Einstein, expanding on the quantum theory, explained this by suggesting that light is composed of tiny bundles of energy, called light quanta or photons. When these photons are of the right energy level, or color, they can dislodge electrons from the atoms of some metals, causing the flow of electric current. Unlike dim blue light, photons of red light are too weak to achieve this. Einstein's theory did not replace the wave theory, which explained the phenomena of diffraction, reflection, refraction, and dispersion, but supplemented it by ascribing particle properties to light.

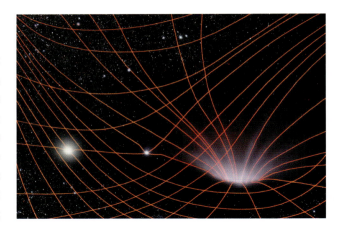

**CONCEPT OF WARPED SPACE**

The idea that space is distorted by gravity is a consequence of Einstein's Theory of Special Relativity. The grid of space is like a thin rubber sheet on which objects of varying weight produce smaller or larger dents. The sun, at bottom left, makes almost no impression. A small, but much denser and more massive, neutron star (lower center) creates a slight distortion. The enormous gravitational pull of a black hole (bottom right) creates a yawning chasm, warping the fabric of space for light years around.

Based on Planck's law that "energy of radiation is proportional to its frequency," Einstein postulated that the greater the frequency of light the greater is the energy of electrons emitted. This led to the invention of laser (light amplification by stimulated emission of radiation) in 1961 by Charles Townes (1915-). In an ordinary bulb there is chaotic motion of the photons of light; it gives white light because it is the mixture of all seven colors of various frequencies or wavelengths. The laser emits light of a single frequency and enormous intensity, practically demonstrating Einstein's theory of spontaneous and stimulated emission of light.

Einstein's Special Theory of Relativity changed our concept of space and time. It held that if, for all frames of reference, the speed of light is constant, and if all natural laws are the same, then both time and motion are found to be relative to the observer. It abolished the concept of absolute time, fundamental to Newtonian laws of mechanics, and pointed the inconsistency in Maxwell's electromagnetic theory. This law began a new paradigm in the world of physics.

The mathematical footnote to the Special Theory of Relativity gave the famous equation: $E = mc^2$, which says that the amount of energy released when an atom is split equals the product of its mass and the square of the velocity of light. The conversion of mass to energy gave the formula for atomic bombs.

## General Theory of Relativity

Einstein published his General Theory of Relativity in 1916, which continues and expands the Special Theory. The Special Theory applies when no accelerations are involved and its effects become noticeable near the speed of light. The General Theory applies when accelerations are involved and in the presence of strong gravitational fields. It unifies Special Relativity and Newton's law of universal gravitation, and explains gravity in terms of the curvature of four-dimensional space-time—the three space dimensions and a fourth time dimension. The assumptions of the General Theory are:

• There is no absolute reference point in the universe;
• Every place is equivalent to observe the universe;
• The center of the universe is wherever we choose it to be;
• The velocity of light in vacuum is independent of the source;
• Nothing can exceed the speed of light in vacuum.

General Relativity is based on the principle of equivalence, which equates the inertial mass appearing in Newton's second law of motion with the gravitational mass in Newton's law of universal gravitation. When a human being is on a disc that is rotating, he experiences centrifugal force that tries to throw him or her on a tangent to the disc, while centripetal force acts toward the center. According to Einstein's principles, these forces are manifestations of gravitation. There is an additional force that is known as coriolis force, which arises due to the rotation of the earth. Einstein observed that the rivers in the northern hemisphere silt soil on the right, while those on the southern hemisphere do the opposite, which he attributed to the coriolis force caused by the earth's rotation.

General Relativity modifies the classical concepts of time, space, light, and the motion of bodies in free fall with its concepts of gravitational time dilation, the gravitational redshift of light, and the gravitational time delay. It is one of the biggest intellectual achievements of the 20th century, providing explanation for phenomena such as black holes, gravitational lenses, gravitational waves, and the expanding universe.

## Quantum Mechanics

In 1912, Niels Bohr (1885-1962), a Danish physicist, joined Rutherford's group studying the structure of atom at

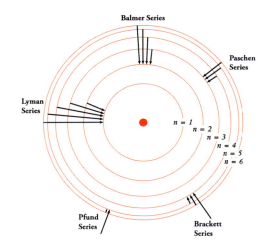

**BOHR'S MODEL OF THE HYDROGEN ATOM**

Niels Bohr's model had electrons occupying discrete (quantized) energy levels (labeled as n, from 1 to 6). The orbitals surround the nucleus. When hydrogen is heated, the electrons absorb energy and rise up one or more levels. When they drop back down again, they emit photons of light, producing an emission spectrum. The series of lines on these emission spectra correspond to transitions back to the same energy levels. The series are: Lyman (n=1), Balmer (n=2), Paschen (n=3), Brackett (n=4), and Pfund (n=5).

## ERWIN SCHRÖDINGER'S CAT EXPERIMENT

To demonstrate the strangeness of quantum mechanics and Heisenberg's uncertainty principle, Schrödinger (left) came up with "Schrodinger's Cat" thought experiment (right) that made him very famous. He imagined a cat placed in a box along with a radioactive device which, like the electron or photon, lives in a dual state—both as a particle and a wave. His premise was that if the device emits one gamma ray it will cause a hammer to fall and open a can of lethal hydrogen cyanide gas which will kill the cat. But there is also the possibility that the radioactive substance will not be a gamma ray and thus the hammer will not fall, which means the cat will survive. In this hypothetical situation, the cat is thought to be both alive and dead until observed. Similarly in quantum mechanics, the unstable particle exists in an intermediate "probabilistic" state until it is observed.

Manchester. He envisioned Planck's quantum theory as guiding the motion of electrons in Rutherford's atom, and modified the structure of atom by introducing different orbits in which electrons spin around the nucleus. According to Bohr, each electron has a fixed amount of energy that corresponds to its fixed orbit. Atoms absorb or emit radiation only when the electrons abruptly jump orbits. The difference in energy levels between orbits divided by Planck's constant h gives the frequency or wavelength of the light emitted.

Bohr's model is similar to the solar system: electrons revolve around the nucleus just as planets revolve around the sun. The attraction between the electrons and the nucleus is similar to the gravitational forces between the planets and the sun. Though Bohr's model described the properties of the hydrogen atom very accurately, it failed to explain the behavior of more complex atoms. Arnold Sommerfield (1868-1951) of Munich, Germany, extended Bohr's work to explain the spectra of light emitted by atoms other than hydrogen and postulated that electrons move in elliptical as well as circular orbits.

Einstein had shown in 1905 that light, a form of electromagnetic waves, is composed of tiny particles called photons. Louis de Broglie (1892-1987), a French physicist, proposed that electrons, like light, also displayed wave-particle dualism. The wave nature of electrons was experimentally established in 1927. In his complementarity principle in 1928, Bohr explained the complementary relation between the wave aspect and the particle aspect of light and electrons.

Meanwhile, scientists in Germany adopted two approaches to formulate quantum mechanics: matrix mechanics, proposed by Werner Heisenberg (1901-1976), Max Born (1882-1970), and Pascual Jordan; and Erwin Schrödinger's (1887-1961) wave mechanics. Matrix mechanics, perceiving electron as a particle with quantum behavior, is based on advanced matrix computations, which introduce discontinuities and quantum jumps. In contrast, wave mechanics, based on the idea of the electron as a wave, entails more familiar concepts and equations. The Schrödinger equation, the fundamental equation of quantum mechanics, describes the wave functions related to the probable occurrence of physical functions and specifies how these waves are altered by external influences. The new mathematical function ψ, the amplitude of Schrödinger's wave, is used to calculate not how the electron moves but what is the probability of finding the electron in a specific place. Born pointed that wave functions determine only the probable position of an electron and not an instantaneous and accurate position and velocity. Werner Heisenberg (1901-1976) spelt it out in his uncertainty principle in 1927, which says the more precisely the position of a particle is determined, the less precisely the velocity is known in this instant, and vice versa. In other words, if we measure one magnitude of a particle—be it its mass, velocity, or position—it causes the other magnitudes to blur. Despite the differences between Schrödinger and Heisenberg, Schrödinger published a proof in 1926 that showed the results of matrix and wave mechanics to be the same. The Copenhagen Interpretation stresses the importance of basing theory on what can be observed and measured experimentally. In 1926, P.A.M. Dirac (1902-1928) combined Heisenberg's matrix mechanics with Schrödinger's wave equations and Born's statistical interpretation in his transformation theory, the first complete mathematical formulation of quantum mechanics.

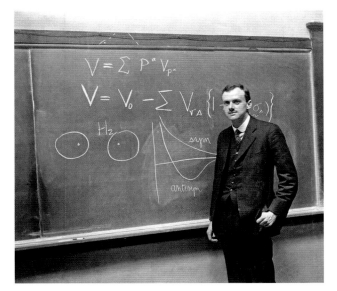

## Quantum Electrodynamics

In 1927 Dirac applied the quantum theory to electromagnetic radiation to obtain a theory about how atoms emit and absorb photons of electromagnetic radiation. This was the beginning of quantum electrodynamics (QED). Dirac further advanced QED in 1928 when he discovered an equation describing the motion of electrons, which reconciled quantum mechanics and the theory of Special Relativity. It was an important step because quantum mechanics provides the most accurate description of matter, and any description involving photons and velocities near the speed of light must involve Special Relativity. Dirac's approach led to the discovery of the process of creation and annihilation of particles. The negative energy solutions of Dirac's relativistic equations showed that electrons have a spin of 1/2 and thus behave like a tiny magnet—as a spinning electron generates a current that, according to Ampère, generates a magnetic field. Dirac also postulated that his solutions of negative energy refer to the existence of antiparticles—particles having the same mass and opposite charge to a electron. This particle—the antielectron, or positron—was accidentally discovered in cosmic rays by Carl Anderson of the US in 1932.

Dirac's quantum field theory explained only one aspect of the electromagnetic interactions—between radiation and matter. During the following years scientists worked on developing a comprehensive theory of QED that explains the interactions of charged particles not only with radiation but also with one another. During the 1930s, physicists such as Heisenberg,

**PAUL DIRAC**

The blackboard displays a quantum mechanical model of the hydrogen molecule. Dirac was a major contributor to the theory of quantum mechanics. His general theory of 1926 was followed, in 1928, by a relativistic theory of quantum mechanics that explained electron spin. He used his theory to predict the existence of antiparticles such as the positron, which was observed in 1932.

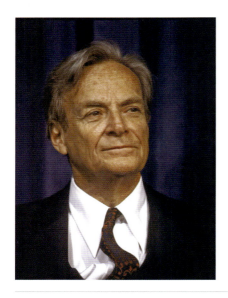

**RICHARD P. FEYNMAN**

In the late 1940s, Feynman developed the theory of quantum electrodynamics, the quantum mechanical view of the interaction of particles subject to the electromagnetic force. His problem-solving tools, including pictorial representations of particle interactions known as Feynman diagrams, influenced many areas of theoretical physics in the second half of the 20th century.

(1906-1979) in the early 1950s. Schwinger and Tomonaga used mathematical, relativistic descriptions of electromagnetic fields. Feynman used descriptions of particle path through space-time, including pictorial representations of particle interactions known as Feynman diagrams. Later, Freeman Dyson (1923-) of the US showed that the two approaches produced the same results, and that Feynman's approach could be derived from the equations of Schwinger and Tomonaga.

## Subatomic Particles

In the 1920s, the nucleus was supposed to be composed of two particles: the proton and the electron. In 1932, James Chadwick (1891-1974), a British physicist, discovered the neutron, a particle with the same mass as proton but no electric charge. The elementary particles seemed firmly established as the proton, neutron, and electron till the discovery of subatomic particles.

Wolfgang Pauli (1900-1958), and J. Robert Oppenheimer (1904-1967) modified the theory of QED to enable it to produce more accurate answers, but the modifications introduced some infinite terms in the equations of QED, which made it even more complicated. For instance, even the mass of a single electron was infinite according to QED because, on the time scales of the uncertainty principle, the electron could continuously emit and absorb virtual photons.

In the late 1940s, American physicists Willis Lamb and Robert Retherford, while studying the energy state of a hydrogen atom, discovered the Lamb shift, which caused physicists such as Oppenheimer, Victor Weisskopf, and Hans Boethe to come up with the theory of renormalization. Acknowledging all possible infinities, renormalization allowed the positive infinities to cancel the negative ones, so the mass and charge of the electron, which are infinite in theory, are then defined to be their measured values. But there was still the problem of consistency with the Special Theory of Relativity, which was resolved by American physicists Richard Feynman (1918-1988) and Julian Schwinger (1918-1994), and the Japanese Tomonaga Shin'ichirō

**ENRICO FERMI, WERNER HEISENBERG, AND WOLFGANG PAULI**
From left to right

Heisenberg gave the matrix theory of quantum mechanics, while Fermi and Pauli contributed to subatomic studies. They were all awarded the Nobel Prize for Physics, in 1932, 1938, and 1945 respectively.

While electrons are fundamental particles with no further constituents, protons and neutrons are both made from smaller particles called quarks. There are six different types of quarks, of which only two are needed to explain protons and neutrons. The proton is made from two up quarks and one down quark; the neutron consists of one up and two down quarks. In 1931, another particle, called the neutrino, was thought of being produced during the decay process. Carl Anderson (1905-1991) discovered the first antiparticle toward the end of 1932 in cosmic rays: the positron, or antielectron. More exotic particles have also been found in cosmic rays, including muons and pions. However, all subatomic particles are not fundamental. Physicists studied these new particles and developed the standard model between 1960 to 1980. This model describes all interactions of subatomic particles, except those due to gravity, and establishes two classes of elementary particles that make up all the matter in the universe and have spins of one-half unit: leptons, which include electrons, muons and neutrinos; and quarks, which come in six varieties and stick together in twos or threes to form heavier particles such as protons, neutrons, and pions. In addition, each of these particles has a corresponding antiparticle with the same mass but opposite electric charge and other properties.

All these particles fall into two categories, based on their statistical behavior: fermions and bosons. Fermions are the quarks and leptons that make up matter; bosons include the photon and other particles associated with forces. Whether a particle is a fermion or a boson is determined by its spin, which is the intrinsic angular momentum of an object. Named after Satyendra Nath Bose of India (1894-1974) and governed by Bose-Einstein statistics, bosons have zero or integer spin—0, 1, 2 and so on. Fermions, named after Fermi, have half integer spin—1/2, 3/2, 5/2 and so on—and obey Fermi-Dirac statistics. The spin has a large effect on the way particles behave. Fermions are the fundamental particles of matter and only created in particle-antiparticle pairs. Bosons act as force carriers that mediate the interactions between particles and can be created and destroyed much more easily. Fermions obey the Pauli Exclusion Principle, formulated by Wolfgang Pauli, which allows only one particle in each quantum state. Bosons do not obey the Exclusion Principle, so any number can occupy the same quantum state. Composite particles such as atoms of Helium-4 (having a nucleus of even mass number) act as boson, whereas Helium-3 (having a nucleus of odd mass number) atoms act as fermions at low energies. Particles of matter acting as bosons can create a great many interesting phenomena at low temperatures, such as Bose-Einstein condensation and superconductivity.

Particles interact by four fundamental forces: gravity, electromagnetism, the strong nuclear force, and the weak nuclear force. These forces, mediated by their individual set of exchange particles, are described by gauge theories.

**TSUNG-DAO LEE (LEFT) AND CHEN NING YANG**

Lee and Yang showed that parity conservation (mirror-image symmetry) is violated in interactions via the weak nuclear force. This work won them the 1957 Nobel Prize for Physics. They also proposed that electron- and muon-type neutrinos are different, predicted the existence of the weak force carrier, the W boson, and indicated the existence of neutral weak currents.

Electromagnetic and gravitation forces operate over long distances and their exchange particles (the photon and the graviton) have no mass. The weak force, responsible for radioactive beta-decay, is mediated by W+, W–, and Z0 over short distances. The strong force, which binds the constituents of the atomic nuclei and is mediated by gluons, also operates over short distances. A quantum field theory for the strong force, called quantum chromodynamics (QCD), was developed in the 1970s.

During the interaction of fundamental particles, it was assumed the overall parity remains the same, or is conserved. This law of the conservation of parity was an intrinsic part of quantum mechanics till Chinese-born physicists Tsung-Dao Lee (1926–) and Chen Ning Yang (1922–) proposed in 1956 that parity is not always conserved. They said that there was no evidence that parity conservation applies to the weak interaction. This was later proved experimentally.

It is believed that in the high energy conditions of the Big Bang, a single superforce governed all particle interactions, which later split into the four forces mentioned above. The ultimate goal of particle physics is to combine all four forces in a unified theory. The electroweak theory, which unifies the electromagnetic and weak nuclear forces, is the first step in this direction. This theory was developed in the late 1960s by Sheldon Glashow (1932–), Abdus Salam (1926-1996), and Steven Weinberg (1933–), and has considerable experimental support. Now the goal for physicists is to discover whether the strong force can be unified with the electroweak force in a grand unified theory. There is evidence that the strengths of the different forces vary with energy in such a way that they converge at extremely high energies. Though the grand unified theory of the strong, weak, and electromagnetic forces may be achieved in a few decades, including gravity may well take much longer, and is the ultimate challenge for physicists.

Other problems that physicists are grappling with include supersymmetry, string theories, and the search for the Higgs particle. This particle is associated with the mechanism that allows the symmetry of the electroweak force to be broken, or hidden, at low energies and that gives the W and Z particles, the carriers of the weak force, their mass. The particle is necessary to electroweak theory because the Higgs mechanism requires

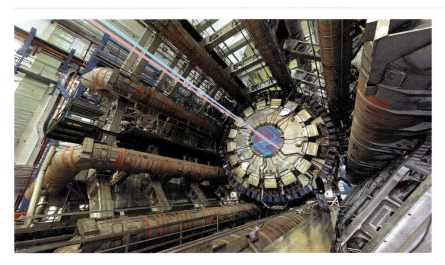

**ATLAS DETECTOR AT CERN**

ATLAS is a particle physics experiment at the Large Hadron Collider at CERN. Starting in Spring 2009, the ATLAS detector will measure the broadest possible range of particles and physical processes that could result from the collision of the proton beams at very high energies. One of the most important goals of ATLAS is to investigate a missing piece of the standard model, the Higgs boson.

a new field to break the symmetry, and according to quantum field theory all fields have particles associated with them. Experiments are being carried out to discover the Higgs particle at CERN.

## Nuclear Fission

Nuclear fission is a process in which the nucleus of a heavy atom is broken apart into fragments, usually two fragments of comparable mass, emitting an enormous amount of energy. This energy is expelled explosively in an atomic bomb.

The history of fission begins with the discovery of the neutron by James Chadwick in 1932. Soon after this, Enrico Fermi and his associates in Italy started an intensive study of the nuclear reactions produced by the bombardment of various elements with this uncharged particle. In 1934, they discovered that at least four different radioactive species, which emitted beta particles, resulted from the bombardment of uranium with slow neutrons. Many radiochemists tried to understand the properties of these new species. In 1939, German physicists Otto Hahn (1879-1968) and Fritz Strassmann (1902-1980), following a clue provided by Irène Joliot-Curie and Pavle Savi in France, proved that these elements were in fact radioisotopes of barium, lanthanum, and other elements in the middle of the periodic table. In most nuclear reactions, an atom changes from a stable form to a radioactive form, or it changes to a slightly heavier or a slightly lighter atom. Copper (element number 29), for example, might change from a stable form to a radioactive form or to zinc (element number 30) or nickel (element number 28). However, the bombardment of uranium gave dramatic results. An atom of uranium (element number 92), when struck by a neutron, broke into two much smaller elements such as krypton (element number 36) and barium (element number 56). The reaction was given the name nuclear fission because of its similarity to the process by

**GERMAN PHYSICISTS LISE MEITNER AND OTTO HAHN**

Meitner and Hahn worked together in Berlin for 30 years. In the 1930s, they worked on the bombardment of uranium with neutrons. In the late 1930s, Meitner's Jewishness became a threat to her safety and she left Germany for Sweden where she built up her research group. In 1939, Hahn in Germany and, a month later, Meitner in Sweden announced that they had obtained nuclear fission for the first time. Hahn was awarded the Nobel Prize for Chemistry in 1944.

which a cell breaks into two parts during the process of cellular fission. German physicist Lise Meitner (1878-1968)—Meitner was a longtime colleague of Hahn who had left Germany due to anti-Jewish persecution—and her nephew Otto Frisch (1904-1979) formulated liquid-drop model of the nucleus to give a qualitative theoretical interpretation of the fission process and called attention to the large energy release that should accompany it.

In 1939, Frédéric Joliot-Curie, Hans von Ha ban, and Lew Kowarski found that several neutrons were emitted in the fission of uranium-235, and this discovery led to the possibility of a self-sustaining chain reaction. Fermi and his coworkers recognized the enormous potential of such a reaction if it could be controlled. On Dec. 2, 1942, they succeeded in doing so, operating the world's first nuclear reactor. Known as a "pile,"

# MANHATTAN PROJECT

In 1939, just before the beginning of World War II, Albert Einstein and other American scientists told President Frank!in D. Roosevelt of efforts in Nazi Germany to purify uranium-235, which could be used to make an atomic bomb. They informed the president about the potential hazard of an uncontrolled fission chain reaction. The US government acted swiftly by setting up the Manhattan Project, a mission to expedite research that would produce an atomic bomb.

The biggest challenge was the production of uranium-235, the essential fissionable component of the postulated bomb, which was very hard to extract at that time. The ratio of conversion from uranium ore to uranium metal is 500:1. Moreover, the extracted metal is 99 per cent uranium-238, which is of no use for an atomic bomb. As uranium-235 and uranium-238 are isotopes, almost identical in their chemical properties, they can only be separated by mechanical methods, and not chemical means. Harold Urey at Coumbia University developed an extraction system that worked on the principle of gaseous diffusion, and Ernest Lawrence at the University of California at Berkeley implemented electromagnetic separation of the two isotopes. Both these processes required large facilities and huge amounts of electric power. A massive plant was constructed at Oak Ridge, Tennessee for this purpose.

**DETONATION OF THE WORLD'S FIRST ATOMIC BOMB AT LOS ALAMOS, USA**

The plutonium device was detonated on July 16, 1945, and had an explosive power equivalent to 10,000 tons of TNT. The flash from the explosion could be seen 250 miles (400 km) away, and the shock wave was felt 50 miles (80 km) away. The details of the explosion were kept secret until after the destruction of Hiroshima on August 6, 1945. The army initially reported the explosion as "an accident at an ammunition dump."

The other fissionable material plutonium-239 occurs naturally only in minute quantities, though it can be produced in reactors through the fission of uranium. This method for the production of plutonium-239 was developed by Arthur Holly Compton at the metallurgical laboratory of the University of Chicago. Following Fermi's success in producing and controlling a fission chain reaction in the reactor pile of uranium-238 at Chicago in December 1942, large-scale plutonium-239 production reactors were built on an isolated tract on the Columbia River, north of Pasco, Washington.

Over a period of six years, from 1939 to 1945, the greatest minds of the world worked together toward creating a working atomic bomb. Along with Robert Oppenheimer, who oversaw the project from conception to completion, famous scientists included David Bohm, Leo Szilard, Eugene Wigner, Otto Frisch, Rudolf Peierls, Felix Bloch, Niels Bohr, Emilio Segre, James Franck, Enrico Fermi, Klaus Fuch, and Edward Teller. USA spent more than $2 billion on this project.

The project reached its goal on July 16, 1945, in a remote part of the New Mexico desert, where the first atomic bomb was tested. The following month, two other atomic bombs produced by the project, the first using uranium-235 and the second using plutonium, were dropped on Hiroshima and Nagasaki, Japan. The bombs killed about 140,000 people in Hiroshima and 80,000 in Nagasaki. Since then, many more have died from afflictions attributed to exposure to radiation released by the bombs.

Nuclear explosions produce both immediate and delayed destructive effects. Blast, thermal radiation, and prompt ionizing radiation cause significant destruction within seconds or minutes of a nuclear blast. The delayed effects, such as radioactive fallout and other environmental effects, inflict damage over an extended period of time, ranging from hours to years.

this device consisted of an array of uranium and graphite blocks and was built on the campus of the University of Chicago.

The Manhattan Project, established soon after the United States entered World War II, developed the atomic bomb, which used uncontrolled fission reactions in either uranium or the artificial element plutonium. Once the war had ended, efforts were made to develop the principles of Fermi's nuclear reactor for large-scale power generation, giving birth to the nuclear power industry.

## CONDENSED MATTER PHYSICS

Condensed matter physics deals with the macroscopic and microscopic physical properties of matter in its "condensed" state, that is, when there are strong interactions between a large number of particles. The most common examples of condensed phases are solids and liquids, which arise from bonding and electromagnetic force between atoms. The extraordinary examples include the superfluid and the Bose-Einstein condensate found in certain atomic systems at very low temperatures, the superconducting phase caused by conduction electrons in certain metallic and ceramic materials, and the ferromagnetic and antiferromagnetic phases of spins on atomic lattices.

Condensed matter physics began as a study of matter in its solid state, especially a study of the electromagnetic, thermodynamic, and structural properties of crystalline solids. It involves the observation of microscopic interactions and the effect they produce at the macroscopic level, so the quantum properties of solids are studied with emphasis on the electronic band structure. Thus conductors, such as metals, are found to contain free conduction electrons that cause electrical and thermal conductivity. Semiconductors and insulators, either crystalline or amorphous, are also studied in this field of physics.

**COLLINS HELIUM CRYOSTAT,** Physics Division of the National Physical Laboratory, Teddington, UK, 1953

This apparatus, designed by US physicist Samuel C. Collins and colleagues in 1946, was designed to liquefy helium. The invention of liquid helium, which forms at extremely low temperatures, opened the door to extensive experimental low-temperature physics.

The most spectacular display of order in macroscopic systems is found in superfluids and superconductors. In these systems, a large fraction of the particles (atoms in superfluids and electron pairs in superconductors) is in the ground state (the state with the lowest energy) and exhibits a coherent behavior at the macroscopic level—the completely frictionless flow in superfluids and the resistance-less flow of electricity in superconductors.

Superfluidity occurs in both of the stable isotopes of helium—helium-3 and helium-4—at temperatures near absolute zero. Superfluidity in helium-4 was discovered in 1938 by Soviet physicist Pyotr Leonidovich Kapitsa (1894-1984). Helium-4 exhibits superfluidity when it is cooled below 2.18 K (-270.97° C), which is known as the lambda ($\lambda$) point. Superfluidity in helium-3 was discovered by American physicists David M. Lee (1931-), Douglas D. Osheroff (1945-), and Robert C. Richardson (1937-). It occurs at temperatures a few thousandths of a degree above absolute zero (-273° C), when helium becomes a superfluid and flows without internal friction found in normal liquids.

Superconductivity is the complete absence of electrical resistance in many solids when they are cooled below a certain temperature. This temperature, called the transition temperature, varies for different materials but generally is below 20 K (-253° C). Superconductivity was first discovered in mercury by the Dutch physicist Heike Kamerlingh Onnes (1853-1923) in 1911. Besides their lack of resistance, another basic property of superconductors is their ability to prevent external magnetic fields from penetrating their interiors. This phenomenon, discovered by the German physicists W. Meissner and R. Ochsenfeld in 1933, made it possible to formulate a theory of the electromagnetic properties of superconductors. For decades a fundamental understanding of the phenomena eluded scientists until a comprehensive theory was developed in 1957 by the American physicists John Bardeen, Leon N. Cooper, and John R. Schrieffer (known as the BCS theory) to explain the behavior of superconducting materials. A similar model was developed by the Russian physicists Lev Davidovich Landau and Vitaly Lazarevich Ginzburg in 1950, which has been useful in understanding the electromagnetic properties of superconductors. The field of superconductivity was revitalized in 1986 when a new class of high-temperature superconductors was discovered by Karl Alex Müller (1927-) and J. Georg Bednorz (1950-). Though the research on these high-temperature superconductors is still going on, they hold great promise for practical applications. The main advantages of devices made from superconductors are low power dissipation, high-speed operation, and high sensitivity. The American physicist Philip Warren Anderson's (1923-) research in semiconductors, superconductivity, and magnetism has made possible the development of inexpensive electronic switching and memory devices in computers.

The phenomenon of superconductivity, in which electrons condense into a new fluid phase in which they can flow without dissipation, is very similar to the superfluid phase found in helium-4 and helium-3. Besides their significance to technology, macroscopic liquid, and solid quantum states are important in astrophysical theories of stellar structure in, for example, neutron stars. The Russian physicist Grigory Volovik (1946-) from the Landau Institute of Theoretical Physics in Moscow won the 2004 Simon Memorial Prize for his pioneering research on the effects of symmetry in superfluids and superconductors, and for extending these concepts to quantum field theory, cosmology, quantum gravity, and particle physics.

### SUPERCONDUCTIVE LEVITATION

According to the Meissner Effect, superconductors have a tendency to repel a magnetic field. Thus a superconductor sample, placed over a magnet, will levitate above the magnet. The height of levitation is governed by the mass of the sample and the strength of the magnetic field.

Recently developed laboratory techniques and theoretical methods have enabled the manufacture of matter exhibiting new types of order. In materials science, materials can be grown atom by atom or, on the scale of micrometers, using colloidal particles as the building blocks. Physicists are now able to create totally new material structures, on scales ranging from nano to mesoscopic dimensions, and then proceed to explore their structural and electronic properties. As a result a whole new field of quantum physics has appeared—the physics of nanostructures—where new phenomena, many with obvious technological applications, are being discovered almost daily.

# CHEMISTRY

# Chapter 1
# FROM ALCHEMY TO CHEMISTRY

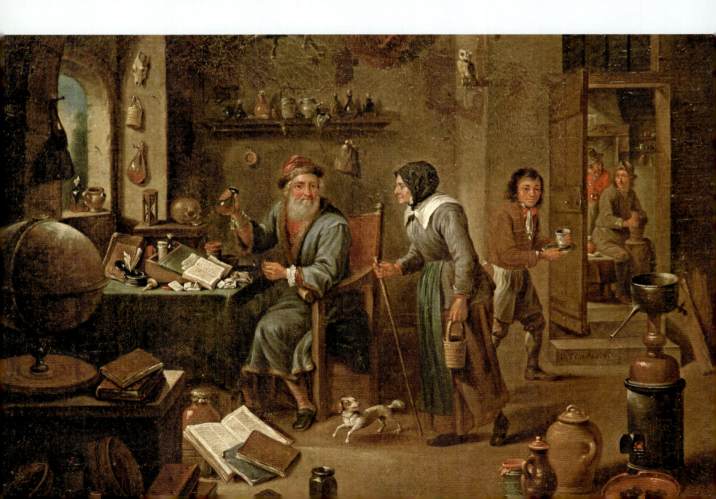

**SYMBOLS OF ALCHEMY,** Title page of *A Very Brief Tract Concerning the Philosophical Stone*, Frankfurt, 1678

Alchemy was the pseudo-scientific predecessor to chemistry, and is best known for its practitioners' hunt for the elixir, which would impart eternal life, and the philosopher's stone, which would turn base metals to gold.

**FACING PAGE:** *The Alchemist's study*, Oil on canvas, Musee des Beaux-Arts, Carcassonne, France

# FIRE, ATOMISM, AND ALCHEMY

The origin of chemistry goes back to the discovery of fire about 9,000 years ago, which was gradually used to harden pottery and extract metals from their ores. The development of metallurgy by ancient cultures allowed the purification of metals and the making of alloys, as well as the exploitation of many minerals and natural substances. Pliny the Elder of Rome (AD 23–79) describes early purification methods and makes acute observations of the state of many minerals in his *Naturalis Historia* of *c.* AD 77. It is believed that the word "chemistry" comes from the Greek word "chemeia"—used in the 4th century BC to designate the art of metal-making, especially the possible change of base metals into gold and silver.

The first theory of matter originated in the 5th century BC in ancient Greece. Empedocles (*c.* 492–432 BC), a Greek philosopher and scientist, said that all matter was made of four elements—fire, air, water, and earth—and that nothing is created or destroyed, but things merely transform from one form to the other as the ratio of these elements change. This theory was challenged by the theory of atomism proposed a few decades later. Lucretius's *De Rerum Natura* (*On the Nature of Things*), written in 50 BC, traces atomism to Leucippus of Miletus (*fl.* 5th century BC) and Democritus of Abdera (*c.* 460–370 BC), who stated that atoms were the most indivisible part of matter. However, this claim lacked scientific evidence. Aristotle (384–322 BC) rejected the existence of atoms and presented his theory of elements on the basis of the theory of Empedocles. He stated that matter was a combination of four elements (earth, air, fire, and water) and four qualities (hot, dry, wet, and cold) in different proportions, and that each of these elements could be transformed into another element through the quality they possessed in common. This led scholars to believe that there exist means for transforming cheaper metals into gold, giving way to alchemy, a quest for transforming base metals into silver or gold. Alchemists searched for the philosopher's stone, which would transform other metals into gold by a mere touch, and

c. **450 BC** Empedocles of Greece states that matter is made of four elements: earth, air, fire, and water.

c. **420 BC** Democritus of Greece declares the atom to be the smallest unit of matter.

**330 BC** Aristotle presents his theory of five elements and rejects the existence of atoms.

**50 BC** Lucretius writes *De Rerum Natura* and supports atomism.

**AD 476** Roman Empire collapses.

**AD 1000–1600** Widespread practice of alchemy in Egypt, China, India, and the Islamic countries.

c. **850** Arab scientist Al Kindi debunks the idea that alchemy can turn base metals to gold.

c. **900** Al Razi of Persia perfects various processes of distillation and extraction.

**1530** Paracelsus, a Swiss-born natural philosopher, explains toxicity.

**1661** Disapproving of alchemy, Robert Boyle compiles the first list of elements and presents his gas law of pressure and volume.

**1700** Johann Becher proposes the phlogiston theory of combustion.

**1754** Joseph Black isolates carbon dioxide.

**1766** Henry Cavendish discovers hydrogen.

**1774** Joseph Priestly discovers oxygen.

**1783** Antoine-Laurent Lavoisier presents the law of conservation of mass.

**1787** The first chemical nomenclature.

**1787** Gay-Lussac publishes the law of expansion of gases.

**1805** John Dalton presents the atomic theory of matter.

**1811** Amedeo Avogadro formulates the law of molecular weight of gases.

the elixir for eternal youth and life. In her 30s, Queen Elizabeth I is reported to have asked an alchemist to find this elixir for her.

## FROM ALCHEMY TO CHEMISTRY

The period from the 1st century to the 17th century AD saw the widespread practice of alchemy in ancient Greece, Egypt, China, India, and the Islamic countries. Though there were regional differences, the practice was based on the basic principle of transmutation—it employed various processes to convert obscure materials into gold.

Alchemists believed there were four spirits (mercury, sulfur, arsenic, and sal ammoniac) and six bodies (gold, silver, copper, tin, lead, and iron), and that the purest bodies (gold and silver) could be obtained by treating the other elements with the four spirits. The metals gold, silver, copper, tin, lead, and iron were known before the rise of alchemy. Mercury and sulfur, both crucial to alchemy, were also known. In a process that involved repeated heatings of various mixtures, alchemists treated metals with corrosive salts, including the chloride of ammonia (also known as sal ammoniac), which came to be known from

**DISTILLATION** *Liber De Arte Distillandi, Simplicia Et Composita* by Hieronymus Brunschwig, 1512

The scientific process of distillation was used by alchemists in the 16th century to manufacture *aqua vitae*, the water of life, better known as brandy. Alchemy developed advanced laboratory techniques, preparing the ground for modern chemistry.

*Chou-i ts'an t'ung ch'i*, a Chinese text of the 2nd century AD. The experiments also led to the discovery of mineral acids—nitric acid, sulfuric acid, and hydrochloric acid.

In the 14th century, cracks began to appear in the theory of alchemy and it diminished in popularity. The discovery of mineral acids allowed alchemists to perform experiments that could recover reactants in their original form. This sparked doubts about the theory of immaterial forms, and supported the earlier theories of tiny atoms that were unchanged by any number of reactions.

At the beginning of the 17th century, Aristotle's theory was still dominant, but experiments by alchemists were beginning to cast doubt on the popular view. These experiments were not based on any scientific rule; there was no proper naming of compounds; the language was esoteric and the methodology bordered on occult. Increasingly, people became skeptical and a need to replace alchemy by a scientific form, where the rules are stated clearly and experiments could be repeated by other people, was felt. Although the goals of alchemy—prolonging life and transmuting base metals into gold—were not accomplished, it laid the foundation for chemistry by developing laboratory techniques, inventing chemical processes like distillation, and discovering new elements, alloys of metals, and mineral acids.

## BEGINNING OF MODERN CHEMISTRY

Some consider medieval Muslims to be the earliest chemists, who introduced precise observations and controlled experimentation into the field, and discovered numerous chemical substances. The most influential Muslim chemists were Geber, Al Kindi, Al Razi, and Al Biruni. The works of Geber (c. 721–815), the Latinized form of Jābir ibn Hayyān, became more widely known in Europe through Latin translations in 14th-century Spain. The work of Indian alchemists, experimenting

**ANTOINE-LAURENT LAVOISIER**

Antoine-Laurent Lavoisier, regarded as the "father of modern chemistry", discovered the law of conservation of mass, overthrew the phlogiston theory and established the role of oxygen in combustion. He co-authored the modern list of chemical elements, among his other achievements.

since antiquity with medicinal properties of plants as part of the science of Ayurveda, also began to draw attention.

In the 13th century, Roger Bacon, Albertus Magnus, and Raymond Lully pondered on the futility of the search for the philosopher's stone and said that alchemy should be directed towards more useful goals. Later, in the 16th century,

Paracelsus (1493–1541), a Swiss natural philosopher, believed that alchemy should be employed in the cure of the sick and used salt, sulfur, and mercury (associated with the alchemical elixir of life) as basic elements for treating diseases. He rejected the four-elements theory and formed a hybrid of alchemy and sciences in what was to be called iatrochemistry. The influences of philosophers such as Sir Francis Bacon (1561–1626) and René Descartes (1596–1650), who believed in removing bias from scientific observations, led to a scientific revolution.

Modern chemistry began to emerge when Robert Boyle made a clear distinction between chemistry and alchemy in 1660. In his *The Skeptical Chymist* (1661), Boyle rejected the earlier theories about the nature of matter and compiled the first list of elements. He elucidated the relationship between the volume and pressure of a gas in an equation known as the Boyle's law.

German chemists Johann Joachim Becher (1635–82) and Georg Ernst Stahl (1660–1734) presented the phlogiston theory in the

## ROBERT BOYLE (1627–1691)

Best known for the gas law that bears his name, Irish-born English philosopher, naturalist and chemist Robert Boyle pioneered modern chemistry through scientific observation, experiment, and analysis. Rejecting the prevalent Aristotelian theory about the composition of matter, Boyle presented the hypothesis that corpuscles, or atoms, were the finest division of matter. In a series of experiments with Robert Hooke (1635–1703), Boyle studied the behavior of gases quantitatively. He investigated the pressure–volume relationship of a gas sample and published the result in 1662 in the form of a law. Boyle's Law states that the pressure of a fixed amount of gas is inversely proportional to the volume of the gas at a constant temperature.

The apparatus used by Boyle was very simple: a bent glass tube with mercury filled to a certain level. The glass tube he used was bent at one-third of its length and sealed on the shorter side. He added a little mercury through the open end of the tube to seal off a quantity of air in the closed end of the tube. Then he measured the volume of the enclosed air for various amounts of added mercury. It was seen that when the unbalanced length of the mercury column increased, the length of the air column, and thus the volume of the air sample, decreased. He concluded that the air volume and the force acting on it were inversely proportional. Later it was established that this relation requires the temperature to be kept constant and that many gases, as well as air, behave in this way. Using the same equipment, Boyle discovered several physical traits of air,

such as its role in combustion, respiration, and the transmission of sound.

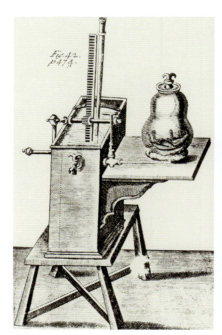

**BOYLE'S AIR PUMP**

Robert Boyle and his collaborators invented this apparatus at the end of the 1650s. A bent handle pumped air out of the bell jar, leaving a near-vacuum. Experiments on candles and animals proved that air was essential for life.

late 1600s, on which the current theory of combustion is based. English chemist Joseph Priestley (1733–1804) discovered in 1774 that oxygen is essential to the burning process. Hydrogen had already been identified by Henry Cavendish (1731–1810) in 1766. Based on these discoveries, the French chemist Antoine-Laurent Lavoisier (1743–1794) formulated the theory of combustion. In 1783, Lavoisier developed the theory of conservation of mass, where he showed that the products of a chemical reaction have the same total mass as the reactants, no matter how much the substances are changed. This demands careful measurements and quantitative observations of chemical phenomena. Lavoisier established the consistent uses of chemical balance, used oxygen to overthrow the phlogiston theory, developed a new system of chemical nomenclature, and made contributions to the modern metric system. He translated the archaic and technical language of chemistry into something that could be easily understood by the largely uneducated masses, and thereby increased the public interest in chemistry. The contributions of Lavoisier led to what is usually called the chemical revolution, marking the beginning of modern chemistry.

The next breakthrough in chemistry came when John Dalton (1766–1844), an English schoolteacher, developed the atomic theory in 1805. He claimed that all elements were made of small particles called atoms and that the atoms of an element were identical. Atoms could not be created, divided or destroyed in a chemical process—they could only combine with atoms of other elements to form a chemical compound. To distinguish atoms of different elements, Dalton set about calculating their relative weights. His work was expanded upon by the Swedish chemist Jöns Jacob Berzelius (1779–1848), who determined the atomic weight of 40 elements and established the chemical formulae for the organic compounds known at that time.

Around the same time, in France, Joseph-Louis Gay-Lussac (1778–1850) was working on his law of the combining volume

**JOHN DALTON,** Historical artwork

Dalton's atomic theory was based on his observations of the behavior of gases. Here Dalton is seen collecting marsh gas (methane), which is formed by rotting vegetation and is trapped under a bed of degrading matter. Disturbing the bed with a pole releases some gas, which can be captured in a jar. Dalton established that gases were made up of combinations of atoms in different quantities.

of gases. In 1805, he established that hydrogen and oxygen combine by volume in the ratio 2:1 to form water, so that $H_2O$, and not Dalton's HO, was the formula for water. As opposed to Dalton's gravimetric approach, Gay-Lussac's approach to the study of the matter was volumetric. It was in 1811 that Italian chemist Amedeo Avogadro (1776–1856) reconciled Dalton's atomic theory with Gay-Lussac's volumetric law in his molecular hypothesis, postulating that equal volumes of different gases at the same temperature and pressure contain the same number of molecules, and that the atoms of some of the gaseous elements are joined together in molecules rather than existing as independent atoms. These theories explained why only half a volume of oxygen combined with a volume of carbon monoxide (CO) to form carbon dioxide ($CO_2$). Each molecule of oxygen has two atoms, and each of these atoms joins one molecule of carbon monoxide: $O_2 + 2CO \rightarrow 2CO_2$. However, Avogadro's molecular hypothesis was not accepted until half a century later, when another Italian chemist, Stanislao Cannizzaro (1826–1910), constructed a logical system of chemistry based on the hypothesis.

# Chapter 2
# A NEW FORM OF CHEMISTRY

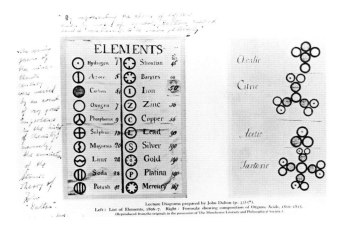

Lecture Diagrams prepared by John Dalton (p. 233(*)).
Left : List of Elements, 1806-7. Right : Formulæ showing composition of Organic Acids, 1810-1813.
(Reproduced from the originals in the possession of The Manchester Literary and Philosophical Society.)

In the 5th century BC, the Greek philosopher Democritus (c. 460-370 BC) put forward the theory that all matter consists of very small, individual particles. He named these particles atomos (meaning indivisible). Although his theory was not accepted by many of his contemporaries (most notably Plato and Aristotle), the notion of 'atomism' endured.

## JOHN DALTON

In 1808, an English scientist and schoolteacher, John Dalton (1766–1844), formulated a precise definition of the indivisible building blocks of matter that we call atoms. Dalton's work marked the beginning of the modern era of chemistry. The hypothesis about the nature of matter on which his atomic theory is based can be summarized as follows:

1: Elements are composed of extremely small particles called atoms. All atoms of an element are identical: they have the same size, mass, and chemical properties. No two elements have the same atoms.

2: Compounds consist of atoms of more than one element. The ratio of the number of atoms of any two elements present in a compound is either an integer or a simple fraction.

3: A chemical reaction involves only the separation, combination, or rearrangement of atoms; it does not result in their creation or destruction.

Dalton's concept of an atom was far more detailed and specific than that of Democritus. His first hypothesis states that the atoms of one element are different from those of another. He had no idea what an atom was really like and therefore made no attempt to describe its structure or composition. But he did realize that the different properties of elements such as hydrogen and oxygen can be explained by assuming that hydrogen atoms are not the same as oxygen atoms.

The second hypothesis suggests that in order to form a compound, we need atoms of the right kinds of elements and a specific number of these atoms. This idea is an extension of a law published in 1799 by a French chemist, Joseph Louis Proust (1754-1826). Proust's law of definite proportions, also called the law of constant proportions, states that different samples of a compound always contain the constituent elements in the same proportion by mass. Thus, if we were to analyze samples of carbon dioxide gas obtained from different sources, we would find in each sample the same ratio (by mass) of carbon to oxygen. And if the ratio of the masses of different elements

**JOSEPH LOUIS PROUST (1754-1826)**

Proust was the son of a chemist, and his early experiments concerned the properties of sugar. In 1797, he formulated the Law of Definite Proportions, also known as Proust's Law. This stated that all compounds contain elements in certain definite proportions, regardless of the mass of the elements and the conditions of their formation.

# HUMPHRY DAVY

A British chemist, Humphry Davy (1778-1829), however, dismissed Dalton's theory as "rather more ingenious than important." Davy was one of the most important figures in 19th-century science. About Dalton's claim that atoms are neither created nor destroyed in a chemical reaction, Davy said, "There is no reason to suppose that any real indestructible principle has yet been discovered."

When Davy was very young his father, a woodcarver, sent him to apprentice with a surgeon in his hometown of Penzance. The apprenticeship enabled Davy to conduct serious chemical experiments and by the time he was 21, he had written *Researches, Chemical and Philosophical*, which led to his appointment to the Royal Institution. During the early part of the 19th century, Davy concluded through his experiments that many common substances were formed by mixing oxygen and metals. In the process, he discovered metals such as potassium, sodium, barium, and strontium, which were not commonly found in their pure state.

in a given compound is fixed, the ratio of the atoms of these elements in the compound must also be constant.

Dalton's second hypothesis supports another important law, the law of multiple proportions: when two elements combine to form more than one compound, the mass of one element that combines with a fixed mass of the other element is in a ratio of small whole numbers. Dalton's theory explains it quite simply: Different compounds made up of the same elements differ in the number of atoms of each element. For example, carbon forms two stable compounds with oxygen, namely, CO (carbon monoxide) and CO2 (carbon dioxide). Thus, the ratio of oxygen in carbon monoxide to oxygen in carbon dioxide is 1:2. This result is consistent with the law of multiple proportions.

Dalton's third hypothesis is another way of stating the law of conservation of mass, or that matter can be rearranged but it cannot be created or destroyed. Because matter is made of atoms that remain unchanged in a chemical reaction, it follows that mass must be conserved as well. Dalton's brilliant insight into the nature of matter was the main stimulus for the rapid progress of chemistry during the 19th century.

Davy's most important investigations were devoted to electrochemistry. Experiments conducted by an Italian physician and physicist, Luigi Galvani (1737-1798), and the invention of the first electric battery, the voltaic pile, created widespread interest in galvanic electricity. William Nicholson (1753–1815), a British chemist, and Anthony Carlisle (1768–1842), a surgeon, used the voltaic pile to obtain hydrogen and oxygen from water. Davy, too, began to experiment with the battery and found that some substances decomposed when he passed an electrical current through them, a process later called electrolysis.

Davy first tried to separate the metals by electrolyzing aqueous solutions of the alkalis, but this yielded only hydrogen gas. He then tried passing current through molten compounds and was

**HUMPHRY DAVY'S MINERS' SAFETY LAMPS**
Invented in 1815, Davy's safety lamps proved a boon to the mining industry, preventing explosions ignited by the flame torches used by miners.

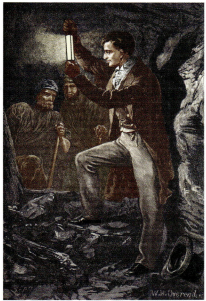

### HUMPHRY DAVY (1778-1829)

British chemist Davy's most important work was in the field of electrochemistry, where he discovered the process of electrolysis. Davy isolated the elements potassium, sodium, calcium, magnesium, barium, and strontium, by passing an electric current through their compounds. He also discovered nitrous oxide, or laughing gas, the gas that was used as the first anesthetic. Davy was knighted in 1812.

able to separate globules of pure metal. In 1807, he separated potassium from molten potash, and sodium from common salt. Through electrolysis, Davy eventually discovered magnesium (named after Magnesia, a district in Thessaly), calcium, strontium, and barium. Several other chemists had also done work on isolating pure elements from compounds, but it was Davy who was able to isolate pure metals and identify them.

In 1810, by a series of experiments, he showed that muriatic acid or marine acid was a compound only of hydrogen and chlorine, and contained no oxygen. For example, he found that two volumes of muriatic acid react with mercury to give calomel and one volume of hydrogen, putting an end to the French chemist Antoine Lavoisier's (1743-1794) theory that oxygen was an essential constituent of acids.

Davy was knighted in 1812 and became one of the most well-known scientists in England. He was also remarkably inventive and is credited with several inventions, including the design of a new arc lamp, a process to desalinate sea water, and a method of fixing zinc plates to the copper-plated hull of ships, thereby increasing the life of the copper plating. This latter method was of utmost importance given the growing importance of maritime trade and warfare in the 19th century.

But perhaps the most important and famous of Davy's inventions was the miner's safety helmet. This was a time when coal was fueling the steam revolution and there seemed to be an insatiable demand for coal. The coal miners used to work in extremely hazardous conditions with frequent explosions and flooding. The major cause of explosions was the methane in mines, which was highly inflammatory and caught fire because of the lamps used by the miners in the dark mines. Davy's lamp, fixed to the helmet, could burn safely in this environment and proved to be a huge boon to the mining industry. Interestingly, Davy did not patent the lamp, an error that led to subsequent false claims by a locomotive engineer, George Stephenson, that he had invented the safety helmet.

In 1827, Davy became seriously ill and the illness was later attributed to his inhalation of many gases over the years. In 1829, he went to live in Rome and died of a heart attack the same year in Switzerland.

Independent of Davy's research, many other scientists produced path-breaking work around that time. Two French chemists, Pierre Louis Dulong (1785–1838) and Alexis Thérèse Petit (1791–1820), proposed their Dulong and Petit's Law (1819) which states that all chemical elements have approximately the same atomic heat; or, the same quantity of heat is needed to heat an atom of all simple bodies to the same extent. Dulong also described the explosive properties of nitrogen trichloride.

In 1811, a French chemist, Bernard Courtois (1777–1838), accidentally discovered iodine when he was observing purple vapors rising from kelp ashes that he had acidified with sulfuric acid and heated. The purple vapors condensed and, on a cold surface, formed dark, shiny crystals. Later, Joseph Gay-Lussac and Humphry Davy proved that the crystals were an element and named it iodine after the Greek word for violet, iodes.

**PIERRE LOUIS DULONG (1785–1838)**

Originally a physician, Dulong discovered in 1811 the explosive nitrogen trichloride, losing an eye and two fingers in the process. But his major work was on the specific heat of atoms.

**BARON JONS JACOB BERZELIUS (1779–1848)**

The Swedish chemist Berzelius produced the first accurate table of atomic masses for 28 elements and reintroduced initial letter symbols to denote elements. He also came up with the concepts of isomerism and catalysis.

## JONS JACOB BERZELIUS

With the isolation and identification of several new elements and compounds, there was a need for a uniform and convenient chemical notation. This problem was solved by a Swede, Jons Jacob Berzelius (1779–1848), who was a doctor. He developed a chemical notation in which elements were represented by letters, typically the first letters of their chemical notation. Thus, oxygen was O while iron was Fe and copper Cu, etc. This is the same system that we use today, except that Berzelius used superscripts for compounds ($H^2O$) while today we use subscripts and write water as $H_2O$.

Berzelius also discovered the law of constant proportions while conducting some experiments for a chemistry book he was writing for his medicine students. The law basically states that inorganic substances are composed of different elements in constant proportions by weight. In discovering that atomic weights are not integer multiples of the weight of hydrogen, Berzelius also disproved Proust's hypothesis that elements are built up from atoms of hydrogen.

He identified the chemical elements silicon, selenium, thorium, cerium, lithium, and vanadium, and coined the terms catalysis, polymer, isomer, allotrope, and protein. Berzelius was the first person to make the distinction between organic and inorganic compounds.

## FRIEDRICH WÖHLER

One of Berzelius' students was a German, Friedrich Wöhler (1800-1882), who was driven by a desire to obtain the finest education in chemistry and went to Sweden to study with Berzelius. Even after returning to Germany, he remained Berzelius's loyal supporter and translated several editions of the Swedish doctor's textbook into German. The University of Göttingen, where Wöhler taught for nearly 50 years, became the most important center for the study of chemistry.

One of the early ideas about chemicals was that they split into two types. Organic chemicals were thought to be found only in living things. It was believed that organic chemicals were different from inorganic ones because they contained a special life force. In 1828, Wöhler accidentally synthesized urea, an organic compound known to occur in living things, from an inorganic substance (ammonium cyanate). Never before had an organic compound been synthesized from inorganic material, and an excited Wöhler wrote to his mentor Berzelius: "I can no longer, so to speak, hold my chemical water and must tell you that I can make urea without needing a kidney, whether of man or dog; the ammonium salt of cyanic acid is urea." It was hard to explain how a life force could enter a compound simply by heating it. Eventually, chemists gave up the life force theory, which was also known as vitalism. Wöhler's experiment did not prove that vitalism was absent; only that the theory was not capable of explaining the result of his experiments.

This opened a new field of research in chemistry, and by the end of the 19th century, scientists were able to synthesize hundreds of organic compounds including mauve, magenta, and other synthetic dyes, as well as the widely used drug aspirin. Today, well over 14 million synthetic and natural organic compounds are known, a number that is far greater than the over 100,000 known inorganic compounds.

In 1825, Wöhler had a dispute with a fellow German chemist, Justus von Liebig (1803-1873), over two substances that had apparently the same composition—cyanic acid and fulminic acid—but very different characteristics: the silver compound of fulminic acid, investigated by Liebig, was explosive, whereas silver cyanate, as Wöhler found, was not. (They amicably

### FRIEDRICH WÖHLER (1800-1882)

Wöhler's accidental synthesis of urea from inorganic ammonium cyanate decisively debunked the age-old 'life force' theory of organic matter.

**LIEBIG'S TEACHING LABORATORY AT GIESSEN UNIVERSITY, 1840**

The German chemist Liebig began teaching at Giessen University in 1824. There he created a laboratory that students could use, and the Giessen University thereafter became the world's premier center for chemistry for over a quarter of a century. Liebig worked in the new field of organic chemistry, perfecting a method of quantitative organic analysis. In later life he became interested in biochemistry, studying the composition of body fluids and experimenting with chemical crop fertilizers.

resolved their dispute and subsequently collaborated in pathbreaking research). Berzelius called these substances 'isomers' and chemists believed that they are defined not only by the number and kind of atoms in the molecule but also by the arrangement of those atoms.

## THE CHEMISTRY OF CARBON

All organic chemicals contain the element carbon, and in view of its importance, a whole science of organic chemistry is dedicated to the study of carbon. The science is called organic because it used to be the study of living organisms (living things consist of carbon compounds), but is now the study of all compounds that contain carbon, except for 'inorganics' such as carbonates and carbon dioxide. More for convenience than any other reason, small molecules like carbon dioxide or carbon monoxide are not counted as organic chemicals. Carbon is different from all other elements because it has the unique ability to form very stable bonds with itself. As a result, there are long chains containing hundreds of thousands of carbon atoms. Organic compounds can be divided into families such as proteins, fats, and sugars.

## FRIEDRICH KEKULE

One of the most important ideas in organic chemistry is that of the carbon atoms joining together to form a ring. In 1865, a Germen chemist, Friedrich August Kekule von Stradonitz (1829-1896), put forward the idea of a ring structure for benzene, an organic liquid with a powerful aroma, after he dreamt of a snake biting its tail. Kekule originally studied architecture but became interested in chemistry when he heard Liebig (of condense flame fame) give evidence in a murder trial. His creative period began with a stay in London as assistant to J. Stenhouse at St. Bartholomew's Hospital.

From 1855 to 1858, Kekule was first a lecturer in Heidelberg, where he debated with J.F.W. von Baeyer, who developed the strain theory of triple bonds in small carbon rings. Kekulelater taught at Ghent and Bonn universities. He was neither a good practical chemist nor an inspiring teacher. His main contributions to chemistry were theoretical and speculative. At the time he began his research, most chemists thought it impossible to determine the structure of molecules as reactions would disturb the structure unpredictably.

In 1858 Kekule postulated that:

• Carbon atoms can combine with one another to form chains of any length and complexity.

• The valency of a carbon atom is always four.

• The study of reaction products can give information about structure.

The 'Structural Theory' idea came to him when he was daydreaming on a bus ride and saw images of carbon atoms dancing before his eyes. When he woke up, he spent the whole night working out the consequence of this idea. He then presented his major paper, 'Notes on some products of substitution of benzene', to the Royal Academy of Belgium, in which he reported his conclusions that the structure of benzene was a closed, hexagonal, six-member ring.

**WILLIAM HENRY PERKIN (1838-1907)**

In 1856, while still a student at the Royal College of Chemistry, the English chemist Perkins accidentally produced mauveine, the first synthetic dye. He set up a factory to manufacture this new dye and his factory marked the beginning of the dye industry. Perkins also synthesized the amino acid glycine. At the age of 36, Perkin gave up dye-making to concentrate on research and developed a general synthesis of aromatic acids (the Perkin reaction).

Kekule was concerned about the unresolved questions in chemistry and, in 1860, initiated the first international congress of chemists at Karlsruhe, where many eminent scientists tried to address questions of nomenclature and definitions of atom, molecule, and equivalency. Finally, in 1862, Kekule published the theory of unsaturated carbon compounds. His oscillation theory of rapidly interchanging double bonds in the benzene ring explained the existence of only distributed derivative in various syntheses. These findings contributed to the synthesis of the aromatics and dye industry in 1872.

## SYNTHESIS OF CARBON COMPOUNDS

Organic compounds that contain the benzene ring structure are called aromatics. The aromatic compound aniline, also called amino benzene, is the starting point for an entire range of vivid dyes called aniline dyes. Organic compounds that are made of chains of carbon atoms with no rings are called aliphatics. The first aniline dye mauveine was accidentally discovered by the English chemist William Henry Perkin (1838–1907) when he was trying to make synthetic quinine. He extracted a purple substance from the mixture he was working on and found that it could dye silk. He called it mauve. Perkin went on to build a dye-works that marked the beginning of the dye industry.

**FRIEDRICH AUGUST KEKULE VON STRADONITZ (1829-1896)**

Kekule's idea of a hexagonal ring structure for benzene came after he dreamt of a snake biting its tail. Similarly, the 'Structural Theory' idea of carbon compounds came to him while daydreaming on a bus ride, when he saw images of carbon atoms dancing before his eyes.

In 1858, Scottish chemist Archibald Scott Couper (1831-1892) and Kekule said that carbon atoms can link directly to one another to form carbon chains. This discovery was based on Kekule's theory proposed in 1857 that carbon is tetravalent. Representing complex organic molecules with many atoms linked in chains was a very challenging task until Couper, in 1858, wrote a paper on salicylic acid, where for the first time he used straight lines between symbols for elements to represent bonds, a system of representation still in place.

Couper, who studied in Glasgow, Edinburgh, Berlin, and Paris, came to chemistry from the study of philosophy and classical languages. He presented his paper on carbon linkage before the French Academy a few weeks after Kekule published his similar paper. Couper had kept his research paper with French chemist Charles-Adolphe Wurtz (1817-1884) in Paris and suffered a nervous breakdown when Wurtz procrastinated in giving the paper to an Academy member for presentation. Couper went back to his Scottish home and never published another scientific paper for the remaining 30 years of his life.

**CHARLES GOODYEAR (1800-1860)**

Goodyear accidentally invented vulcanized rubber when he spilled a hot mixture of sulfur and rubber. He patented the process in 1844 and, in 1852, a French company was licensed by Goodyear to make shoes. But Goodyear did not earn much profit, and died in debt.

**ARCHIBALD SCOTT COUPER (1831-1892)**

Scottish organic chemist Couper independently discovered the theory of organic structure and proposed the first ring structure of any compound. But just before his paper could be presented to the French Academy, Kekule came up with a similar theory and took the credit. Couper lived his later years in obscurity.

Apart from the dye and pharmaceutical industry, organic chemistry also contributed to several other areas. In the 18th century, natural rubber was discovered by the French in the jungles of South America. Although it was useful because of its unique properties, its use was restricted because it melted in hot weather and froze too fast.

In 1838, Charles Goodyear (1800-1860) came to know that sulfur was being used in the manufacture of rubber. One day he accidentally spilled the mixture of sulfur and rubber on a hot stove and the result was vulcanized rubber, which could be used in industry in many forms. He patented the process in 1844, and in 1852, a French company was licensed by Goodyear to make shoes. In 1855, the French Emperor gave Goodyear the Grand Medal of Honor and decorated him with the cross of the Legion of Honor. Finally in 1898, almost four decades after his death, the Goodyear Tire and Rubber Company was founded.

# THE STRUCTURE OF THE ATOM

On the basis of Dalton's atomic theory, an atom can be defined as the basic unit of an element that can enter into a chemical combination. Dalton imagined an atom as something not only extremely small but also indivisible. However, a series of investigations that began in the 1850s and extended into the 20th century revealed that atoms actually possess an internal structure: they are made up of even smaller particles called subatomic particles. This research led to the discovery of three such particles: electrons, protons, and neutrons.

## THE ELECTRON

In the 1890s, many scientists began to study radiation, or the emission and transmission of energy through space in the form of waves. Their research contributed greatly to the understanding of atomic structure. One device used to investigate this phenomenon was the cathode ray tube, the forerunner of the television tube. It is a glass tube from which most of the air has been sucked out. When the two metal plates in the tube are connected to a high-voltage source, the negatively charged plate, called the cathode, emits invisible rays, which are called cathode rays. The cathode rays are drawn to the positively charged plate, the anode, where they pass through a hole and continue traveling to the other end of the tube, which is coated with a chemical that emits a strong fluorescence when the cathode rays hit it.

An English physicist, Joseph John Thomson (1856-1940), used a cathode ray tube to determine the ratio of electric charge to the mass of the particles that made up the cathode rays. Cathode rays were known to be negatively charged since they were attracted towards the positively charged plate. The problem was to quantify the charge of the cathode rays. Thomson could not measure this but instead determined the ratio of the charge to the mass of the charged particles comprising the cathode rays (electrons). He came up with a negative number (- 1.76 x $10^8$ C/g), where C stands for coulomb, the unit of electric charge.

Thereafter, in a series of experiments carried out between 1908 and 1917, an American physicist, Robert Andrews Millikan (1868-1953), succeeded in measuring the precise charge of the electron. His work also proved that the charge on each electron was exactly the same. In his experiment, Millikan examined the motion of single tiny drops of oil that picked up static charge from ions in the air. He suspended the charged drops in air by applying an electric field and followed their motions through a microscope. Using his knowledge of electrostatics, Millikan found the charge of an electron to be 1.6022 x $10^{-19}$ C. From this data and Thompson's value, he calculated the mass of an electron to be 9.10 x $10^{-28}$ g, which is an exceedingly small mass.

**ROBERT ANDREWS MILLIKAN (1868-1953)**

Through 1908 and 1917, American physicist Millikan followed the trajectory of tiny drops of oil when suspended in an electric field and determined the mass of an electron to be the infinitesimally small figure of 9.10 x $10^{-28}$ g. His work, carried out over a series of complex experiments, also proved that the charge on each electron is the same.

THE NEW ROENTGEN PHOTOGRAPHY.
"LOOK PLEASANT, PLEASE."

## RADIOACTIVITY

In 1895, the German physicist Wilhelm Röntgen (1845-1923) noticed that cathode rays caused glass and metals to emit very unusual rays. This highly energetic radiation penetrated matter, darkened covered photographic plates, and caused a variety of substances to fluoresce. Since these rays could not be deflected by a magnet, they could not contain charged particles, as cathode rays do. Röntgen called them X-rays because their nature was not known. X-rays revolutionized medical diagnostics since they provided a way to see images of bones and internal organs. In fact, the first X-ray picture taken, on the night when Röntgen discovered the unusual properties of the radiation, was of his wife's hand!

Apart from their well known application in medicine, X-rays are also used extensively in engineering and physics. In engineering, they are used primarily for determining faults and cracks, etc, in structures, especially metallic structures. In physics, they find use in crystallography, where the crystal structure is determined by the pattern on the spectrum created by the crystal when it is bombarded with an X-ray beam. This technique was discovered in the early 20th century by the father-son duo, W.H. Bragg and W.L. Bragg. X-rays are now also used to create laser beams of very high energy that are used in fusion reactions.

Not long after Röntgen's discovery, Antoine-Henri Becquerel (1852-1908), a physics professor in Paris, began to study the fluorescent properties of substances. Purely by accident, he discovered that exposing thickly wrapped photographic plates to a certain uranium compound caused them to darken even without the stimulation of cathode rays. Like X-rays, the rays from the uranium compound were highly energetic and could not be deflected by a magnet, but they differed from X-rays in that they arose spontaneously.

One of Becquerel's students, Marie Curie (1867-1934) suggested the name radioactivity to describe this spontaneous emission of particles and/or radiation. Since then, any element that spontaneously emits radiation is said to be radioactive. In 1903, Marie Curie and her husband Pierre were awarded the Nobel Prize for Physics "in recognition of the extraordinary services they have rendered by their joint researches on the radiation phenomena discovered by Professor Henri Becquerel."

Radioactive substances such as uranium produce three types of rays or particles when they decay, or break down. These rays are usually deadly for living matter. Apart from severe burns, they can cause mutations in cells. The rays are called alpha, beta, and gamma. Alpha particles or rays are the heaviest of the three and are positively charged. They are the nuclei of the element helium. Beta particles are basically electrons and are negatively charged. Gamma rays are not charged and are a form of electromagnetic radiation like light and radio waves, though significantly more energetic than either of these.

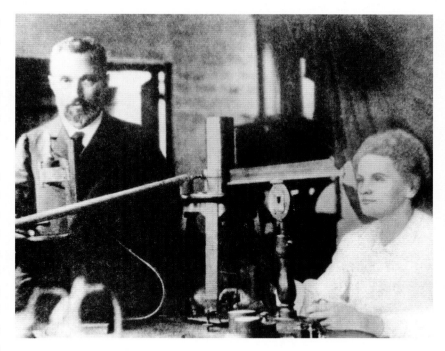

**MARIE AND PIERRE CURIE (1867–1934 AND 1859–1906)**

This photograph of the Curies was taken soon after their marriage in Paris in 1895. Marie Curie (born Marya Sklodowska) began studying radioactivity in uranium shortly after its discovery by Becquerel in 1896. She and Pierre detected two new elements, polonium (named after Marie's native country, Poland) and radium, both highly radioactive. The Curies were awarded the 1903 Nobel Prize for physics for their work. Marie also won the 1911 chemistry prize, five years after Pierre's death in a street accident.

## THE PROTON AND THE NUCLEUS

By the early 1900s, two features of atoms had become clear: they contain electrons, and they are electrically neutral. To maintain electric neutrality, an atom must contain an equal number of positive and negative charges. Therefore, Thomson proposed that an atom could be thought of as a uniform, positive sphere of matter in which electrons are embedded like resins in a cake. This so-called 'plum-pudding' model was the accepted theory for a number of years.

In 1910, New Zealand physicist Ernest Rutherford (1871–1937), who had studied with Thomson at Cambridge University, decided to use alpha particles to probe the structure of atoms. Together with his German associate Hans Geiger (1882–1945) and an undergraduate student named Ernest Marsden, Rutherford carried out a series of experiments using very thin foils of gold and other metals as targets for particles from radium, a radioactive source. They observed that the majority of particles penetrated the foil without getting deflected or with only a slight deflection. But every now and then, an alpha

## ERNEST RUTHERFORD (1871-1937)

New Zealand-born physicist Sir Ernest Rutherford discovered that radioactive decay occurs by successive disintegrations of atoms. In 1911, he proposed a model for the atom in which the positively charged protons were concentrated in a small region, the nucleus, which was surrounded by negatively charged electrons.

within the atom. Whenever an alpha particle came close to a nucleus in the scattering experiment, it experienced a large repulsive force and therefore a large deflection. Moreover, an alpha particle traveling directly toward a nucleus would have its direction reversed.

The positively charged particles in the nucleus are called protons. In separate experiments, it was found that each proton carries the same quantity of charge as an electron and has a mass of $1.67262 \times 10^{-24}$ g, about 1,840 times the mass of the electron.

The picture of the atom thus emerging now was as follows: The atom consists of a positively charged nucleus and also has negatively charged electrons. The mass of a nucleus constitutes most of the mass of the entire atom, but the nucleus occupies only about 1/1,013 of the volume of the atom. Atomic (and molecular) dimensions are expressed in terms of the SI unit (the international system of units) called the picometer (pm), where 1 pm = $1 \times 10^{-12}$ m.

particle was scattered (or deflected) at a large angle. In some instances, the alpha particle actually bounced back in the direction from which it had come. This was a most surprising finding, for in Thomson's model the positive particles should have passed through the foil with a small deflection. To quote Rutherford's initial reaction when told of this discovery, "It was as incredible as if you had fired a 15-inch shell at a piece of tissue paper and it came back and hit you."

Rutherford was later able to explain the results of the scattering experiment in terms of a new model for the atom: most of the atom must be empty space. This explains why the majority of alpha particles passed through the gold foil with little or no deflection. The atom's positive charge, Rutherford proposed, was concentrated in the nucleus, which is a dense central core

### RUTHERFORD'S APPARATUS

Apparatus with which Ernest Rutherford first observed in 1919 the mutation of nitrogen atoms into oxygen atoms, after they collided with alpha particles generated from a source inside the tube.

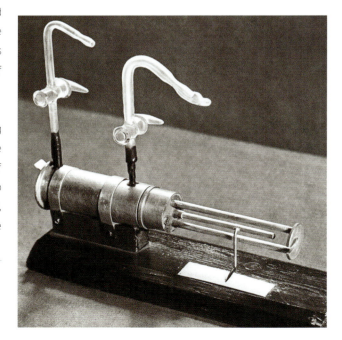

A typical atomic radius is about 100pm, whereas the radius of an atomic nucleus is only about $5 \times 10^{-3}$ pm. To put the relative sizes of an atom and its nucleus in perspective, if an atom were the size of the Houston Astrodome, the volume of its nucleus would be about the size of a small marble. While the protons are confined to the nucleus of the atom, the electrons are conceived of as being spread out about the nucleus at some distance from it.

## THE NEUTRON

Rutherford's model of atomic structure left one major problem unsolved. It was known that hydrogen, the simplest atom, contains only one proton and the helium atom has two protons. Therefore, the ratio of the mass of a helium atom to that of a hydrogen atom should be 2:1 (because electrons are so much lighter than protons, their contribution to atomic mass can be ignored). In reality, however, the mass ratio is 4:1. Rutherford and others postulated that there must be another type of subatomic particle in the atomic nucleus and the proof was provided by another English physicist, James Chadwick (1891–1974), in 1932.

When Chadwick bombarded a thin sheet of beryllium with alpha particles, the metal emitted a very high-energy radiation similar to gamma rays. Subsequent experiments showed that the rays actually consisted of a third type of subatomic particle, which Chadwick named neutron, because it proved to be an electrically neutral particle having a mass slightly greater than that of the proton. The mystery of the mass ratio could now be explained. In the helium nucleus there are two protons and two neutrons, but in the hydrogen nucleus there is only one proton and no neutrons; therefore, the ratio is 4:1.

## AMEDEO AVOGADRO

In 1811, the Italian chemist Amedeo Avogadro (1776–1856) propounded the hypothesis that has since been named Avogadro's Law. The hypothesis stated that equal volumes of ideal gases, at the same temperature and pressure, contained an equal number of molecules. The hypothesis is an extremely important one, since it basically asserted that the size of molecules does not determine the number of molecules contained in gases and implied that the gas constant (the ratio of the product of pressure and volume to the product of the temperature and the amount of the gas) is the same for all gases. The number of molecules in one mole of any substance (a mole is the amount of any substance which contains the same number of molecules as 12 gm of Carbon-12) is a constant (6.02 x 1,023) and is named Avogadro's number in his honor.

By the time of Avogadro's death, although scientists had a fair understanding of chemical processes and structure, there were still several issues being debated—for instance, how does one define a molecule or an atom? Or, how does one determine quantitatively atomic properties like atomic weight?

The Italian chemist Stanislao Cannizzaro (1826-1910) used Avogadro's hypothesis to understand the fundamental structure of chemical compounds and also the atomic weights of various atoms. Born in Palermo, Sicily, Cannizzaro attended medical school where he became interested in chemistry. In 1848, he joined the antimonarchical revolution and after its failure, was forced to flee to Paris. He resumed his study of chemistry in Paris and worked out the reaction to explain the self oxidation and self reduction of aldehydes, a reaction now known as 'Cannizzaro reaction'.

## MENDELEEV'S PERIODIC TABLE

By this time, with the proliferation of chemical elements (there were by now more than 60 distinct elements known), chemists were trying to find a system of classification. Classification of elements was not just a bookkeeping exercise with some kind of logical placement of elements in categories; it was hoped that it would also give some idea of what kind of elements could be discovered. Two chemists, Julius Lothar Meyer (1830-1895) and Russian Dimitri Ivan Mendeleev (1834-1907), were among those trying to put the chaos in some kind of order. The problem was really a lack of consensus on how to determine accurate atomic weights of the elements.

Meyer started off by arranging 28 elements into 6 families, with the principle that elements of a family were similar in their chemical and physical properties. Meyer used the valence of an element as a key indicator of its properties. Valence is a number that represents the combining power of an element. For instance, the valence of oxygen is 2 as it takes two atoms of hydrogen to combine with 1 atom of oxygen to form water. In Meyer's table, the families were organized by decreasing valence, beginning with carbon, which has a valence of 4.

### AMEDEO AVOGADRO (1776-1856)

Lorenzo Romano Amedeo Carlo Avogadro, Conte di Quaragne e Ceretto, laid down a fundamental law of chemistry that equal volumes of gases, at the same temperature and pressure, contain an equal number of molecules. The number of molecules in one mole of any substance is a constant (6.02 x 1,023) and is named Avogadro's number in his honor.

Mendeleev was also working on some form of classification of the elements. Appointed to the chair of chemistry at the University of St. Petersburg in 1867, Mendeleev came out with the Periodic Law in 1869. His classification was remarkable since it contained all known elements into one table, unlike Meyer's which only had 28 elements. As with most revolutionary ideas, Mendeleev's classification initially met with a lot of resistance from chemists. However, with the discovery of new elements that had been predicted by the Periodic Law, most chemists became converts. The Periodic Law or the Periodic Table of Mendeleev soon became the bedrock of chemistry.

**ROBERT BUNSEN (1811-1899)**

Bunsen, a German chemist, discovered the technique of spectroscopy for analyzing the unique spectrum of light produced by each element. He also invented the Bunsen burner, used in almost all school chemistry labs.

**LEFT:** An 1882 spectroscope. The image in the middle shows the spectrum of various elements.

## THE SPECTROSCOPE

One of the most important techniques in chemistry to identify and measure various chemical substances, spectroscopy, was discovered in the 19th century. The credit for its discovery goes to Robert Bunsen (1811-1899), a German chemist, who is more well known for creating the Bunsen burner, a piece of equipment that has become standard equipment in all chemistry laboratories. Bunsen invented the spectroscope in 1859 along with his colleague at the University of Heidelberg, Gustav Kirchhoff (1824-1887). Bunsen also made several improvements to the existing chemical batteries and invented the Bunsen battery. He also discovered the antidote to arsenic poisoning.

After the development of the Bunsen burner, Kirchoff used a prism to analyze the light given off by various chemical substances when they were placed in a Bunsen burner. The prism could split the different wavelengths of light and this information could be used to identify the substances. Thus, for instance, when common salt is placed on a Bunsen burner, it gives off a characteristic yellow light because of the presence of sodium in common salt (sodium chloride). The prism, together with a telescope, could be used as a useful tool for studying different kinds of light given off by various substances, and this formed the spectroscope. A spectroscope has since been used heavily in chemistry to identify and differentiate between substances. In fact, in 1860, Bunsen and Kirchhoff discovered two new alkali metals, cesium, and rubidium, with the help of a spectroscope.

Thus, at the turn of the century, the study of chemistry was at a threshold—the classification of elements and chemical compounds was almost complete. Organic chemistry, with its proliferation of new compounds, was providing a variety of new chemicals, leading to entirely new industries like dyes and drugs. The tools for studying compounds and their properties were also developed. The discovery of atomic structure and radioactivity provided entirely new insights into the nature of matter at its most fundamental level. But developments in this field had to wait until the next century.

# Chapter 3
# THE GIANT LEAP

**NYLON SPINNERS**

Workers at a nylon factory in 1953, from Pontypool, South Wales. Threads of nylon are being spun onto spools. Nylon was the first artificial fiber to be used extensively and was first produced on February 28, 1935, by Wallace Carothers (1896–1937) at the chemical company E.I. du Pont de Nemours.

**FACING PAGE:** A 1912 artwork depicting a modern steel factory. Harry Brearley (1871–1942) created the first-ever stainless steel in 1913 while he was trying to find erosion-resistant steel for his client, a small arms manufacturer.

The 19th century saw some of the most revolutionary advances in technological and scientific achievement. The industrial revolution, the increase in speed of transport and communications, the developments in medicine and agriculture, all led to a dramatic change in the lives of a large percentage of the world's population. As we have seen, chemistry played a major role in these transformations. But the pace of change experienced in the previous two centuries was nothing compared to what the world would see in the brave new 20th century.

The use of chemistry in manufacturing things of utility for humanity, which had started with the dye industry in the previous century, saw many innovations, some of which have become an essential part of our lives: plastics, synthetic textiles, stainless steel, etc. Apart from things visible in our daily lives, chemistry also helped develop products which, though not commonly visible, are of immense utility to humanity—fertilizers, chemicals and, in a controversial way, more powerful explosives.

## FIBERS AND PLASTICS

The production of artificial fibers actually started in the 19th century with the development of 'artificial silk' from nitrocellulose. However, the real breakthrough came with the production of rayon from cellulose. Rayon was not a synthetic fiber, since it was made from a natural product, cellulose. However, its production from cellulose involved extensive processing and it came to be seen and treated like a synthetic fiber. Production of rayon textiles started in 1899 in Germany and the fabric was found to be extremely useful in hot and humid climates.

The most popular synthetic fiber was, however, Nylon, which was first produced at the American chemical company, E.I. du Pont de Nemours. This fiber was synthesized by adopting some of the techniques that had been used to make plastics from petroleum products. Nylon was a very versatile fabric and, because of its light weight and resilience, found use in textiles, hosiery as well as parachutes and other products.

If artificial fibers revolutionized the textile industry, the introduction of plastics proved to be a boon for a variety of industries. Though plastics are environmentally harmful, their proliferation in our daily lives makes it difficult to imagine a world free of plastics. Bakelite was the first plastic, made by the Belgian chemist L.H. Baekeland (1863–1944) in 1909. The discovery of Bakelite happened by accident when Baekeland

**LEO BAEKELAND (1863-1944)**

The Belgian chemist Baekeland's invention of Bakelite, an inexpensive, nonflammable and versatile plastic, ushered in the 'plastic revolution', which is currently cause for environmental concern. Baekeland also invented the 'Velox' photographic paper, which he sold to Kodak.

was the German organic chemist Hermann Staudinger (1881–1965), who did extensive work in the field of polymers. Among his many achievements was a new process to synthesize isoprene, which is used in making synthetic rubber. He also proposed that polymers were basically giant molecules that were held by ordinary chemical bonds, a view radically different from the prevailing view of polymers being disordered combinations of small molecules. Staudinger's work on polymers ultimately led to the establishment of the synthetic polymer industry, manufacturing synthetic rubber, moldable plastics, adhesives, and many more.

## PETROCHEMICALS, GLASS, AND RUBBER

The popularity of the internal combustion engine that was fed with gasoline led to a greater dependence on petroleum. Petroleum refining techniques improved during this period and this allowed the emergence of petrochemicals and plastics. Thermal cracking, discovered in 1913, was a major advance and was used extensively during this period. In this process, distilled

discovered a compound of carbolic acid and formaldehyde. When he tried to reheat the solidified compound, he discovered it would not melt, no matter how high the temperature. This heat-resistant and insulating property of Bakelite made it ideal for a variety of purposes like electrical switches, though initially it was used for all kinds of purposes, including the manufacture of billiard balls and artificial jewelry.

The development of plastics was a result of the improvements in petroleum refining and a better understanding of the chemistry of petroleum. Plastics had been made from organic chemicals in the 19th century (celluloid, Parkesine, etc.). After the discovery of thermal cracking and later of catalytic cracking, the plastics industry exploded. Many new plastics, mostly made from petroleum, were introduced for a variety of uses. Paints, pipes, films, and a host of other applications were where plastics replaced other materials. They were also used extensively in the construction industry.

Plastics are basically examples of polymers, giant molecules whose different composition and structure is responsible for their distinctive properties. The founder of polymer chemistry

**HERMANN STAUDINGER (1881-1965)**

German organic chemist and founder of polymer chemistry, Staudinger initially worked on the aroma agents of coffee. In the 1920s he began to study rubber and was the first to suggest that it was made of large, strand-like molecules, or polymers. He was awarded the Nobel Prize for chemistry in 1953.

Petroleum refining techniques improved rapidly during the early 20th century. Thermal cracking, discovered in 1913, was a major advance. In this process, distilled petroleum was heated under pressure, leading to its 'cracking' the heavier molecules present in the petroleum distillate into lighter ones like petrol.

petroleum was heated under pressure, leading to its 'cracking' the heavier molecules present in the petroleum distillate into lighter ones like petrol. Catalytic cracking, a process in which zeolite was used as a catalyst (a chemical that makes a chemical reaction possible without taking part in the reaction), was discovered in 1936 and led to increased yields of petrol.

The 20th century also saw the development of several other new synthetic materials that proved to be immensely useful for humankind. Fiberglass, a form of synthetic glass, was first developed by Games Slayter (1896–1964), an American inventor in the 1930s. Heat- and shock-resistant glassware, sold under the brand name Pyrex by Corning Inc., was developed in 1915. Pyrex underwent many changes and found use not only in kitchenware but also in laboratory glassware.

During World War I, the supplies of natural rubber from South America and the Far East were disrupted. This led to the development of an industrial process to make synthetic rubber by German scientists, though at least one form of synthetic rubber had been made in the 1890s in the laboratory. After the war, the use of synthetic rubber declined. In 1933, scientists invented a new form of rubber called Buna S, which soon became hugely successful.

## STEEL AND ALUMINIUM

Steel had been known to humans for several centuries as an alloy of carbon and iron that was remarkably strong and had properties that made it extremely useful for producing weapons. The famous Damascus steel was used to make swords. In the mid-19th century, the Bessemer converter was invented. It provided the first inexpensive method for the mass production of steel from molten pig iron. But the search for steel resistant to weather and rust continued without success.

In 1872, the English company Messers Woods and Clark patented an alloy of iron, chromium and tungsten that was resistant to acid and weather. This could be considered the first patent for what came to be known as stainless steel. After this, several kinds of alloys of iron and chromium as well as iron and other elements were made to produce different kinds of steels. However, it was only in 1913 that Harry Brearley (1871–1942) created the first-ever stainless steel, while he was

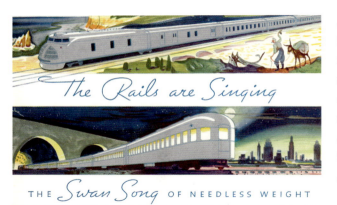

**ADVERTISEMENT FOR ALUMINUM, 1935**

Aluminum was a lightweight alternative to steel and copper. It was used in electrical appliances, in buildings, and in aircraft construction.

trying to find erosion-resistant steel for his client, a small arms manufacturer. The steel that Brearley created had 12.8 per cent chromium and 0.24 per cent carbon. It was resistant to chemical attacks and, hence, to rust.

Stainless steel became widely used over time in cutlery, weapons, and industry. Subsequent researches produced various kinds of specialty steels with varying quantities of other elements, such as vanadium and molybdenum. The consumption of steel in an economy became an indicator of the level of its economic progress, since everything from buildings to machinery to weapons needed steel.

Aluminum had been discovered and isolated in the 19th century but it could not be produced in large enough quantities to be useful. This changed when the process of extraction using electricity was discovered in 1886. With the huge increase in electricity generation in the early 20th century, extraction of aluminum became feasible. Aluminum was a lightweight alternative to steel and copper in many applications. It found use in the electrical industry, in buildings, in aircraft construction, and in household appliances.

## THE ALLOTROPES OF CARBON

The element carbon is not only the basis of all life, but also a remarkably versatile element. Firstly, because of its tetravalent nature, carbon can form an enormous variety of compounds. Secondly, it occurs in allotropic forms, like amorphous carbon, graphite, and diamond, that have very different physical properties. Graphite, for example, is soft enough to be used in pencils, while diamond is the hardest known substance.

Diamonds, besides being prized as gems, are also used in industry, where they are used in drilling rigs and other equipment. Ever since the late 18th century, when diamond was recognized as an allotropic form of carbon, people have tried unsuccessfully to produce synthetic diamonds. In 1941, the General Electric Company (GE) embarked on a project to make artificial diamonds. After many false starts, in 1955, Tracy Hall, a scientist at GE, announced the first commercially successful synthesis of an artificial diamond.

For an element that has been so extensively studied over centuries, carbon sprang another surprise in 1985 when a new allotrope of carbon was discovered. Fullerene, named in honour of Buckminster Fuller (1895–1983), the American architect and polymath, is a 60-atom carbon molecule. The molecule is a perfect sphere of 60 carbon atoms, which are joined in

12 pentagons and 20 hexagons, like the patches on a soccer ball. Fullerenes were first discovered when carbon electrodes produced an electric arc in a helium-rich atmosphere.

Subsequently, other fullerenes with 28, 32, 50, 70, and 76 carbon atoms were also discovered. Because of its rich variety, chemists feel that fullerene chemistry might prove as important as the organic chemistry based on the benzene ring. An amazing variety of new molecules based on fullerenes have been synthesized. Carbon nanotubes, which are cylinders of carbon atoms in a hexagonal arrangement capped by fullerene at the ends, are already being used in a variety of applications. They are many times tougher than other known fibers and, hence, can be used in medical and defense industries. The fuller implications of fullerene chemistry are still being understood. Fullerenes are now being used in industries such as lubricants, pharmaceuticals, and electronic displays.

**FERDINAND FREDERICK HENRO MOISSAN (1852-1907)**

Moissan pictured with his electric furnace. Born in Paris, he was an artificial diamond maker. With his experiments, he managed to study the production of carbon in its three varieties. He was awarded the Nobel Prize for chemistry in 1906.

## NITROGEN-BASED FERTILIZERS AND EXPLOSIVES

The tremendous increase in the human population in the 20th century was in part made possible by the use of artificial fertilizers, which could boost agricultural productivity. The big breakthrough in this field came in 1908 when Friedrich Wilhelm Ostwald (1853–1932), a German chemist, invented the process to manufacture nitric acid which can be used as a base for nitrogenous fertilizers. Although the basic process was well known for several years, the problem was the scarcity of ammonia.

It was Fritz Haber (1868–1934), a German chemist, who in 1908 developed a process to make ammonia gas from the nitrogen in the air. The process involved distillation of nitrogen from liquid air and using hydrogen made from the Bosch process, which produces hydrogen from steam by using methane and a nickel oxide catalyst. Once hydrogen and nitrogen are available, they are mixed at high temperature and pressure in the presence of a magnetite catalyst to produce ammonia.

The processes for production of ammonia and nitric acid had a dramatic impact on two industries—agricultural production, which could now boom because of nitrogenous fertilizers, and armaments, which required these compounds for production of explosives. In fact, it is conjectured that had it not been for the Haber-Bosch process for the production of ammonia, Germany would not have entered the First World War.

Industrial chemistry, or chemical engineering, became one of the biggest industries in the 20th century. The production of inorganic chemicals, organic chemicals, petrochemicals, agrochemicals, and explosives became multi-billion dollar businesses and there is scarcely an area of human existence untouched by these industrially produced products.

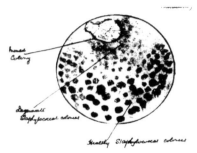

## ANTIBIOTICS AND VITAMINS

It was not only chemical products and materials that saw a huge change in the 20th century. Pharmaceuticals radically changed the way people lived. Not only was mortality reduced substantially by the introduction of new and more powerful drugs to conquer life-threatening ailments, morbidity also declined considerably. In 1928, the biochemist Alexander Fleming (1881–1955) discovered penicillin in his laboratory by accident. As part of an experiment to grow some colonies of bacterial staphylococcus in his laboratory, Fleming discovered that a mould had formed on one of the plates. He found that the mould was inhibiting bacterial growth.

This was a remarkable discovery and further experiments revealed that the fungus, *Penicillium notatum*, could kill many kinds of bacteria and could be administered to laboratory animals without danger. It was 10 years later that Howard Florey (1898–1968) and Ernst Chain (1906–1979) isolated the substance responsible for the anti-bacterial action in the mould and thus was born the first antibiotic, penicillin.

World War II saw the use of antibiotics for the first time in treating infections among the wounded. The war effort provided the impetus for the mass production of the drug, which came to be referred to as the 'wonder drug' because it saved thousands of lives during the war. Subsequent to the isolation of penicillin, many other antibiotics were synthesized against different kinds of bacterial strains. The development of broad-spectrum antibiotics led to a drastic downfall in fatalities due to infectious diseases, and increased the life expectancy in most countries.

This was also the time when many of the vitamins were identified. Deficiency diseases like scurvy and beriberi had been known for a long time and, over the years, specific preventive agents were discovered. For instance, it had been known for more than a century that eating citrus fruit prevents scurvy. But the identification of the specific agents whose deficiency caused the disease was missing.

In 1912, the Polish chemist Kazimierz Funk (1884–1967) identified one such complex of chemicals. He called this complex Vitamine. Over the next few years, many scientists worked to discover the chemicals responsible for preventing diseases and these were all given the generic name Vitamin, with the specific chemical being identified by a suffix. By 1943, almost all the known vitamins had been identified and their importance to the human body understood.

# MORE WONDER DRUGS

In 1922, the first insulin injection was given to a diabetic with stupendous results. Frederick Banting (1891–1941), a Canadian doctor, had managed to extract the chemical from the pancreas of a calf. Over the next few years, insulin extracted from animals became the standard treatment for diabetes and another major disease became treatable.

Apart from antibiotics, many other drugs were discovered or synthesized after World War II. The oral contraceptive pill developed in the 1950s was a revolutionary breakthrough in reproductive medicine. Cortisone, various drugs against hypertension, and other drugs became available in the 1950s and 1960s. The synthesis of steroids, in particular of cortisone,

was an important step as steroids held the promise of being the new wonder drug, much like penicillin was in the 1940s. Diazepam, a tranquilizer, was synthesized in 1960 and soon became a blockbuster, becoming the most prescribed drug in history.

Improved techniques for synthesis and also better methods of mass production meant that drugs could go from the laboratory to the field in a much shorter time span. This also meant that the pharmaceutical industry became a much larger and hugely capital-intensive enterprise, with the costs of developing a new drug and bringing it to the market escalating to more than a billion dollars. Thus, only huge pharmaceutical companies could enter this field. Increasingly, the focus shifted from diseases that impact a majority of the poor in the developing world, to lifestyle diseases that are more prevalent in the developed world.

An example of this is the development of drugs against diseases like heart ailments and psychological depression. With the isolation of mevastatin, an inhibitor of cholesterol production in the body, there has been lots of research in synthesizing other inhibitors to reduce the quantity of Low Density Lipoprotein (LDL) in the body. LDL builds up in the body causing the blockage of arteries. The condition is called arteriosclerosis and is a major cause of mortality in the developed world and, increasingly, in the developing world. Many other drugs have subsequently been made that lower LDL levels and these are among the best selling drugs in the world.

**FREDERICK BANTING (1891-1941) AND HIS ASSISTANT CHARLES BEST (1899-1978)**

When Canadian physiologist Frederick Banting (right) won the 1923 Nobel Prize for medicine, he gave half the prize money to his assistant Charles Best, saying Best had played an equal role in the discovery of insulin. This picture shows Banting and Best standing with the first dog kept alive by insulin injections following the removal of its pancreas.

**THE SPLITTING OF URANIUM BY NEUTRONS**

Otto Hahn (left) with his colleagues Fritz Strassmann and Professor Haber at a museum in Munich, reconstructing the experiment demonstrating the splitting of uranium by neutrons.

**OTTO HAHN AND FRITZ STRASSMANN'S WORKBENCH**

The equipment used by Otto Hahn (1879-1968) and Fritz Strassmann (1902-1980) in December 1938 to provide the first chemical evidence of nuclear fission products after bombarding uranium with neutrons.

# QUANTUM MECHANICS AND CHEMISTRY

The 20th century not only saw enormous progress in applications of chemistry like drugs and plastics, but also in basic chemistry. The stage was set with the development of quantum mechanics in the early part of the century. The discovery of radioactivity and the isolation of new elements had already opened the window to the nature of matter at the sub-atomic level. Quantum mechanics, developed by Heisenberg, Schrodinger, Dirac, and others in the first two decades of the century, had finally provided a comprehensive picture of the nature and behavior of matter at the subatomic level.

In 1932, the discovery of the neutron by James Chadwick proved to be a turning point in the understanding of the nuclear structure. This was because, unlike the proton, the neutron was neutral and would not be repelled by the positively charged nucleus. Many scientists were engaged in studying the reaction products of bombarding neutrons on nuclei of different elements. Lise Meitner (1878–1968), Otto Hahn (1879–1968), and Fritz Strassmann (1902–1980) studied the effect of neutrons on uranium and found that it could produce barium, a much lighter element. It was later found that what actually happened was that neutron bombardment produced transuranic elements heavier than uranium, which decayed radioactively to produce barium. This discovery proved to be crucial to the development of nuclear weapons. Hahn received the Noble Prize for chemistry in 1944 for his work on the fission of heavy atomic nuclei.

Many transuranic elements, that is elements heavier than uranium, have been discovered since then. In all, 118 artificially produced elements have been identified. Most of these elements are produced in collisions between heavy ions and decay radioactively within seconds or less. Apart from plutonium and neptunium, none of the other elements are

**WILLARD FRANK LIBBY (1908-1980)**

In the late 1940s, Libby led the University of Chicago team that developed radiocarbon dating using the radioactive isotope carbon-14. Libby also worked on the Manhattan Project during World War II, helping to enrich the uranium used in atomic bombs.

**RIGHT:** Carbon dating. All living material contains a radioactive isotope of carbon, carbon-14 (14C), and a stable isotope, carbon-12 (12C) at a known ratio. When the tissue dies, the amount of 12C remains constant, but 14C decays radioactively. Measuring the amount of 14C compared to 12C in a sample indicates how long ago the tissue died.

found naturally. Elements with atomic number greater than 103, i.e. beginning with rutherfordium (atomic number 104), are called super-heavy atoms. These elements are created in quantities on the atomic scale for a few seconds. An element having more than the atomic weight of 118 hasn't been created, though there have been unconfirmed reports of the production of an element with atomic weight of 123.

An extremely important application of radioactivity is radiocarbon dating, which was developed by Williard Libby (1908–1980) in 1947. The technique is used to date artifacts based on the unstable isotope of carbon, carbon 14, which is found in trace quantities in the atmosphere. By measuring the ratio of carbon 14 to the carbon 12 isotope, which is found in all living organisms, Libby was able to calculate the time that had elapsed since the organism died. Radiocarbon dating has become an invaluable tool for archaeologists, geologists, and anthropologists; techniques for using other radioactive elements like potassium have also been developed. Libby won the Noble Prize for chemistry in 1960 for this seminal discovery.

Quantum mechanics provided a framework to understand the atom and its interactions, which is of central importance in chemistry. However, the application of these principles to real molecules was not particularly simple because of the level of complexity of all but the simplest atoms. Advances in mathematical techniques and approximation methods eventually enabled the successful application of quantum mechanics to chemistry.

Quantum mechanics is important for understanding how individual atoms combine covalently to form chemicals or molecules. The application of quantum mechanics to chemistry is known as quantum chemistry. In principle, quantum mechanics can mathematically describe most of chemistry. It can provide quantitative insight into ionic and covalent bonding processes by showing which molecules are energetically favorable to which others and by how much. Most of the calculations performed in computational chemistry, a branch of chemistry that uses computers to assist in solving chemical problems, rely on quantum mechanics.

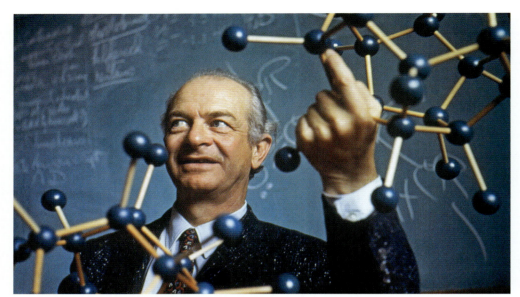

# LINUS PAULING

The person who almost single-handedly achieved the fusion of quantum mechanics and chemistry was the brilliant chemist Linus Pauling (1901–1994). Pauling was a professor at Caltech when he, influenced by European scientists who had worked on quantum mechanics, started applying quantum mechanics to chemistry.

Pauling's early work grappled with the nature of the chemical bond. In a sense, all of chemistry is possible only because atoms bond with each other or with atoms of different kinds. Pauling tried to understand the nature of this bonding. He introduced the idea of orbital hybridization, now one of the central concepts in chemistry. He also worked on the relationship between ionic bonds (in which electrons are transferred between atoms, like in sodium chloride) and covalent bonds (in which electrons are shared between atoms, like in methane). Interestingly, Pauling showed that these two concepts are really the extreme cases and most bonding in actual molecules almost always falls between these two extremes. He formulated the concept of electro-negativity to determine the extent of ionic bonding.

Pauling also studied the structure of aromatic hydrocarbons (like benzene, whose structure had been determined by Kekule in the previous century). Pauling applied the methods of quantum mechanics to the various atoms comprising the benzene molecule and determined that the actual structure was a superposition of single and double bonds, which later came to be known as resonance.

The work of Pauling on the chemical bond culminated in his famous textbook, *The Nature of the Chemical Bond*, which was published in 1939. It is considered by many scientists as the most influential work in chemistry. It placed chemistry, which was until now without a solid theoretical underpinning, on a strong theoretical footing. Pauling received the Nobel Prize for chemistry in 1954 for this work.

Apart from his seminal work on the chemical bond, Pauling also worked on the nature of biological molecules. This was the time, the 1930s, when there was great excitement in the field of biology, especially over the large molecules that play an important role in human biology. The structure of the hemoglobin molecule, which carries oxygen in blood, was still unclear, and Pauling showed that the structure changes when it picks up an oxygen atom or loses it to the body cells.

Pauling then tried to determine the structure of proteins, which are long chains of simpler molecules called amino acids. The structure of the protein determines its functionality and is crucial to their understanding. Pauling used X-ray crystallography pictures of protein molecules but these turned out to be inaccurate. In 1948, Pauling was ill with influenza and to keep himself busy he built paper models of the structure of proteins by linking amino acids in various ways. This was when he hit upon the alpha helix, which is a cylindrical arrangement of amino acids, coiled up and connected with hydrogen bonds. The alpha helix was experimentally verified and Pauling's ideas of applying structural chemistry to biological molecules were widely accepted. Pauling's lock-and-key theory of enzyme action has been used in various situations and has led to many insights in genetics and immunology.

After his work on proteins, Pauling started working on the model for DNA, the fundamental molecule of life contained in the cell nucleus and which carries the entire genetic information of an organism. He was hampered in this work by the lack of good quality X-ray diffraction photographs of DNA. Making good X-ray diffraction photographs of biological molecules was a complicated affair; unlike inorganic materials, it is difficult to make crystals of biological molecules. Pauling initially proposed that the DNA molecule had a triple helical structure, which turned out to be erroneous.

## THE CHEMISTRY OF DNA

Several laboratories in the world were trying to unravel the structure of DNA in the mid 20th century. Maurice Wilkins (1916–2004) and Rosalind Franklin (1920–1958), working at King's College in London, had taken high-quality X-ray photographs of the DNA molecule and were presenting these photographs at a conference that Pauling was going to attend. However, Pauling's passport was impounded because of his work on disarmament just before the conference and he couldn't enter the United States. Consequently, he could not attend the conference and hence had no knowledge about this unpublished work.

As it turned out, two young scientists at Cavendish Laboratory, James Watson (b. 1928) and Francis Crick (1916–2004) were also trying to unravel the DNA's structure around the same time. Interestingly, Watson was working on a completely different project, that of purifying myoglobin (a protein much

**MAURICE WILKINS (1916-2004)**

Wilkin won the Nobel Prize for his technique of X-ray diffraction of DNA, which he developed at King's College, London. Before that, during World War II, he developed improved radar screens at Birmingham, and then worked on isotope separation at the Manhattan Project (to develop the nuclear bomb) at the University of California, Berkeley.

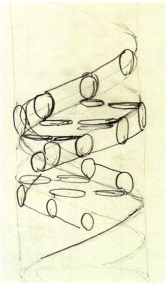

**JAMES WATSON (B. 1928) AND FRANCIS CRICK (1916-2004)**

Watson and Crick with their model of part of a DNA molecule in 1953. The discovery of the double helix structure of DNA was the 20th century's greatest breakthrough in the molecular sciences.

**RIGHT:** Francis Crick's original sketch of the structure of DNA, made in 1953.

like hemoglobin but simpler). Crick, a physicist, had worked on mines during World War II and had come to Cambridge in his later years. Watson and Crick had access to not only the existing information about the chemical structure of the DNA molecule, but also to the X-ray diffraction data of Wilkins and Franklin. In

**ROSALIND FRANKLIN (1920-58)**

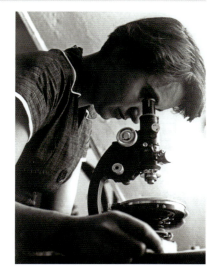

X-ray images of DNA taken by Franklin and her colleague at London's King's College, Maurice Wilkins, played a crucial role in the discovery of the double helix structure of DNA. In 1962, Maurice Wilkins was given the Nobel Prize, along with Watson and Crick. Franklin had died of cancer by then, and was thus ineligible for the award.

1953, they finally worked out the double helical structure of DNA where two helixes were wound around each other like a spiral staircase, with the rungs of the staircase consisting of paired chemical groups of atoms. Franklin died of cancer at the young age of 37, and her contribution to the discovery of DNA's structure was not sufficiently appreciated until Watson wrote his personal account, *The Double Helix*. Watson, Crick, and Wilkins were awarded the Nobel Prize for medicine in 1962.

## RNA AND THE GENETIC CODE

Though DNA was the primary genetic molecule, the translation of the genetic information into protein formation depended on another molecule, ribonucleic acid (RNA). The structure of RNA was deciphered by Severo Ochoa (1905–1993) and it turned out to be very similar to the DNA. The genetic information in the DNA (in terms of the arrangement of the bases) is transcribed to the RNA by enzymes and read in the cell to synthesize proteins. The genetic information resides inside the nucleus while the protein synthesis takes place in the cytoplasm in the membrane.

Once the basic structure of the DNA and RNA molecules was understood, the genetic code needed to be cracked. This was the code that would unravel the mystery of how the sequence of nucleotides in the DNA molecule translates into the sequence of amino acids in a protein. Various scientists succeeded in doing so in the early 1960s.

The English biochemist Frederick Sanger (b. 1918) determined the complete amino acid sequence of insulin in 1955. This was a remarkable discovery since it showed that proteins have a definite structure. Sanger first produced short fragments of insulin by treating an insulin solution with the trypsin enzyme. The short fragments were then applied to some filter paper onto which each fragment left a distinctive mark. These marks could be reassembled into the complete structure of insulin, which represented a precise sequence of amino acids in a well defined structure. Sanger won the Nobel Prize for chemistry for this work in 1958.

Sanger won his second Nobel Prize for chemistry in 1980. Working on ways to speed up DNA sequencing, Sanger in 1975 developed the chain termination method of DNA sequencing, also known as the Dideoxy termination method or the Sanger method. The technique proved to be of immense importance in the Human Genome Project.

## POLYMERASE CHAIN REACTION

Sanger's development of the chain termination method catalyzed the field of molecular biology. Another new technique that played a critical role was Polymerase Chain Reaction (PCR), developed by the American biochemist Kary Mullis (b. 1944) in 1983. This technique allowed a single strand of DNA to produce millions of copies of itself in a relatively short time. PCR proved to be of enormous use in genetic manipulation and in the biotechnology industry. It is inconceivable that the rapid

**FREDERICK SANGER (B. 1918)**

British biochemist Frederick Sanger is one of only four persons (and the only one alive) who have won the Nobel Prize twice in their field. He determined the amino acid structure of the hormone insulin in 1955.

progress made in biotechnology would have been possible without the PCR method for replication. Mullis won the Noble Prize for chemistry in 1993 for his contribution.

A major application of PCR is in forensic science. Typically, small quantities of body fluids or parts (hair, nails, skin) are recovered from crime sites and these can provide vital clues in criminal investigations. However, the material available is usually too little to carry out the various tests. To overcome this, the tiny amount of DNA is run through a PCR and many copies of the original material are thus obtained.

Another powerful technique used in forensic science is DNA fingerprinting or DNA profiling. The basic idea behind this technique, which is used extensively in forensics, is that though more than 99.9 per cent of the DNA is common to all human beings, the rest is different in each individual. This small unique part of the genetic material provides an essentially unique identification of the human being. A small quantity of DNA from an individual is obtained and magnified using PCR. The resulting DNA is then analyzed to find the unique, non-common part of the genetic material that identifies the individual. Sir Alec Jeffreys (b. 1950), a British scientist, developed this technique in 1985.

## 'INORGANIC' LIFE

In the same year that Watson and Crick discovered the structure of DNA, a pathbreaking experiment was performed at the University of Chicago. Harold Urey (1893–1981) and Stanley Miller (1930–2007) showed how complex, organic molecules could have formed from inorganic molecules in the early years of the earth's existence. They took a flask with water, methane, ammonia and hydrogen—substances that could have been present in the primordial atmosphere of earth. They heated the solution and passed electrical sparks through it to simulate lightning flashes. The process was repeated for several days after which the residues were analyzed. To their surprise, they found that complex organic molecules, which are required for life, were present in the mixture. These included amino acids (precursors to proteins), sugars, lipids and some building blocks of nucleic acids. The Urey-Miller experiment proved a hypothesis of several biologists that life originated on the primordial earth from inorganic molecules.

Living beings metabolize, i.e., take in food and produce energy. The chemistry of metabolism was not understood until the 1930s. It was Hans Krebs (1900–1981), a German-born British biochemist, who discovered the citric acid cycle that has come to be known as Krebs' cycle. This cycle is a series of reactions occurring in all organisms that use oxygen to generate energy from carbohydrates, fats and proteins. The Krebs' cycle is of central importance in understanding how living beings are able to convert food into energy and waste products.

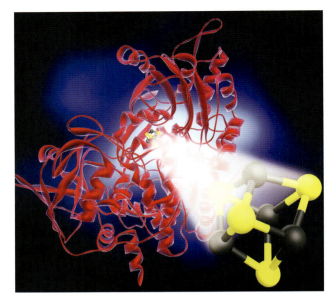

**METABOLIC ENZYME**

Computer artwork of aconitase, an enzyme involved in the Krebs' cycle. The chemistry of metabolism was explained by Hans Krebs (1900–1981), a German-born British biochemist. The Krebs' cycle is of central importance in understanding how living beings convert food into energy and waste.

While metabolism is an essential prerequisite for life, the availability of food equally critical. The photosynthetic process in plants is responsible for converting the sun's energy, together with water and carbon dioxide, to carbohydrates that are used as food by all living organisms. It was known by the end of the 18th century that green plants take in carbon dioxide and emit oxygen in sunlight. It was also known that water is also used by plants for this purpose. However, the details of the process were still a mystery.

The detailed chemical process underlying photosynthesis was understood only in the second half of the 20th century. It was the American chemist Melvin Calvin (1911–1997), along with Andrew Benson and James Bassham, who provided the complete explanation of carbon assimilation in plants. This had profound implications for agriculture.

# NEW BREAKTHROUGHS

The enormous progress in chemistry in the 20th century would not have been possible without the development of tools and techniques that allowed chemists to undertake detailed investigations of matter. Among the most important tool for analytical chemistry is the Nuclear Magnetic Resonance (NMR) spectroscope. Initially proposed by the physicist Isidor Rabi (1898–1988) in 1938, Edward Purcell (1912–1997) and Felix Bloch (1905–1983) used the technique on solids and liquids in the 1940s. After this, several refinements in the technique were made by various scientists, among them being the Swiss chemist Robert Ernst (b. 1933), who developed techniques for high resolution NMR spectroscopy that made NMR a basic tool for analytical chemistry. Ernst's technique allowed two-dimensional exploration of large molecules, something that had not been possible with NMR. This led to NMR being extensively used to determine the structure of large biological molecules. Ernst was given the Nobel Prize for Chemistry in 1991.

The 20th century, thus, was the time when scientists were not only able to understand matter at the most basic level, but also able to comprehend the mechanics of life. It was also a time when the knowledge of chemistry was used to produce materials and processes that enhanced the quality of life. Without petroleum refining, the huge growth in automobile and air travel would have been inconceivable. Without chemical fertilizers, our planet could not have been able to support the enormous increase in human population. Without drugs, human morbidity and mortality would not have been reduced to the lowest levels in history. Without plastics, cement, stainless steel, and a host of other substances, life as we know it would be unthinkable. This huge progress in both fundamental and applied chemistry laid the foundations for the even more dramatic advances of the 21st century.

**NOBEL PRIZE WINNERS AT REUNION MEETING**

Nobel Prize-winning physicists Werner Heisenberg (L), Paul Adrian Dirac (2R), Edward M. Purcell (3L), Gustav Hertz (R), Emilio Segre (5R), Isidor I. Rabi (6R), and Lennart Graff Bernadotte (C), at a reunion in Europe. The work of these physicists was instrumental in shaping the science of atomic and sub-atomic chemistry. The structure of molecules was seen in a new light after the structure and properties of atoms was established by scientists like Heisenberg and Dirac, and through the techniques of atomic analysis invented by scientists like Rabi and Purcell.

**NMR CHEMICAL SPECTROSCOPY**

The nuclear magnetic resonance (NMR) spectrometer, now the basic tool for analytical chemistry, was developed by the Swiss chemist Robert Ernst (b. 1933). In this picture, the spectrum of the sample under investigation appears as the series of peaks on the screen.

# Chapter 4
# THE CHEMICAL CENTURY

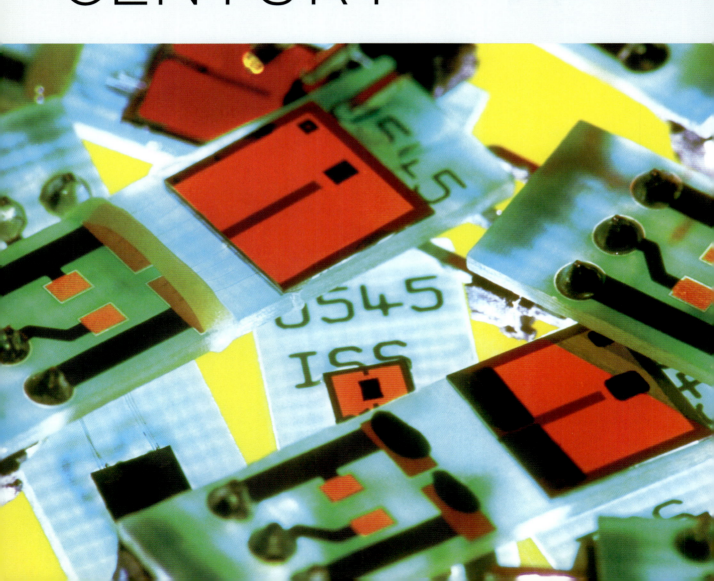

If the tremendous progress made in various branches of chemistry in the last few decades of the 20th century and the first few years of the 21st century is any indication, the 21st century will see revolutionary developments in chemistry. Development of new materials, genomics and biotechnology, nanomaterials and nanoscience, and environmental chemistry are some of the areas receiving the most attention of researchers .

Ever since the first ceramics were created many millennia ago, human beings have developed materials that can be used for a variety of purposes. We have already seen that the 20th century saw the emergence of many new materials and the modification of natural ones. Plastics, artificial fibers, different kinds of steels, semiconductors, and nanomaterials were among the many new materials that came into use in the 20th century.

## CONDUCTIVE POLYMERS

One of the major developments was the development of polymers that could conduct electricity. Polymers are molecules that form long, repetitive chains. The nature of the polymer is dependent not only upon the composition of the chain, but also its structural arrangement. Plastics are among the best known polymers. Ordinarily, plastics are insulators, i.e., they do not conduct electricity. This is because, for electrical conduction, we need free electrons, or electrons that are not bound to any atom but can move around the substance. In the presence of electrical voltage, these free electrons flow and constitute an electrical current.

Although compounds like polyaniline had been known to have electrical properties for almost two centuries, these were never investigated in any great detail. It was not until 1963 that some workers reported finding high conductivity in oxidized iodine-doped polypyrole, a material derived from polyacetylene. Doping is the process by which small amounts of impurities are added to a material to drastically change its properties. The most well-known example of doping is found in semiconductors, where materials like silicon are doped to change their electrical conduction properties.

The next major breakthrough was an actual demonstration of an organic polymer being used as a switch. This happened in 1974 when John McGinness and his collaborators built a device with melanin. Melanin is a mixture of several polymers like polyacetylene, polyaniline, etc. This new switch created quite a sensation, as it added a new dimension to the technological potential of organic polymers.

In 1977, conductive polymers caught the world's attention when A.J. Heeger, A. Macdiarmid, and H. Shirakawa discovered that polyacetylene, doped with iodine, could conduct electricity. This opened up a whole array of possibilities, because manipulating polymers is significantly easier than manipulating metals. Also, since these polymers are much like plastics, one could think of combining them with conventional plastics and create a product that possesses the qualities of both the materials. For instance, combining polymers with plastics that are flexible and yet tough, will give us a material that can conduct electricity as well as a metal wire and yet be as flexible as plastic. Such a material could be used in display panels or touch screens, devices that are remarkably rugged, yet flexible.

Conductive polymers also have the advantage that their conductivity is so high that even with very low voltages they can produce a good amount of light. This has made possible their use in solar panels, flat panel displays, and other such devices. In fact, conducting polymers used as light emitting devices are one of the fastest growing industries and have revolutionized lighting design. Heeger, Macdiarmid, and Shirakawa were recognized for this work with the Nobel Prize for chemistry in 2000.

DOPA melanin, a conductive polymer, is naturally found in many tissues of mammals, and is used for electrical conduction as well as transduction, which is a process that converts energy signals from one form to the other, for instance from light to sound. In the human body, the polymer plays several key functions. One example is its role in the hearing process, where it converts sound vibrations into electrical impulses to be transmitted to the brain. Scientists are developing sensor technologies using conducting polymers for a variety of biological applications, including to speed up the determination of gene sequences, a process known as rapid gene sequencing.

## SOLAR POWER

The growing concern about global warming and greenhouse effect has led to a global effort to reduce fossil fuel consumption. However, the energy demand continues to grow at a very fast pace, especially in the developing countries. Environmental concerns, coupled with the enormous rise in oil prices in the first few years of the 21st century, greatly increased the development of renewable sources of energy like solar power, wind power, and geothermal energy.

Solar power, or the technique of converting the energy received from the Sun into useful and usable energy, has become increasingly important. The development of new materials and technology is being encouraged in the developed world through government incentives. It is not as if we do not use the enormous energy delivered by sunlight to the earth; in fact, our whole existence depends on this source of energy as

**NOBEL LAUREATES, 2000**

The 2000 Nobel Prize ceremony at the Concert Hall in Stockholm, Sweden. Winners of the prize for chemistry were Professor Hideki Shirakawa (center) and, on the left, Alan G. MacDiarmid and Alan J. Heeger (extreme left).

### INTERSOLAR CONFERENCE

Solar panels at the Intersolar North America conference 2008. Solar energy is a major new industry in the developed world, boosted by tax cuts and government subsidies.

### LEFT: SOLAR POWER REFLECTORS

Each dish focuses radiation from the sun onto a thermoelectric generator. A computer steers the dishes to ensure they face the sun.

our food (photosynthesized by plants) and our fuel (organic matter decomposing into fossil fuels) are ultimately derived from solar energy. But the main thrust of the new solar power industry is to efficiently convert this energy into electricity.

The basic principle behind converting solar energy to electrical energy, the photoelectric effect, has been well understood since the beginning of the 20th century. Light falling on certain materials generates an electrical current, and since solar energy is potentially unlimited and the process does not produce any greenhouse emissions, such electricity could solve the energy crisis as well as the problem of global warming.

However, the problem is the low efficiency and, hence, the high cost of producing electric power by this means. Initially, solar cell efficiency, which is the fraction of solar energy falling on the solar cell that converts it into electrical energy, was typically around 1 per cent. Developments in material science over the past century improved this to around 5 per cent, which was still

too low for solar power to be economical except in locations where the conventional grid-supplied electricity cannot reach.

The standard solar cell, called a photovoltaic cell, is made from crystalline silicon. Large arrays of such cells are assembled and used to generate a reasonable amount of electric power. However, in the past few years, there has been a tremendous improvement in the efficiency of solar cells because of the development of new materials and techniques. Thin films of materials like cadmium, tellurium, and amorphous silicon (which is significantly cheaper than the crystalline form) are used and new techniques enable the casting of the thin wafers, unlike in the past when they were cut from a bigger block.

With all these advances, the efficiency of commercially available photovoltaic cells has risen to around 12-15 per cent, making them more economical. However, in laboratory settings, efficiencies of even 40 per cent have been reported. The increased research in the field has been largely inspired by

systems. Even now, fuel cells are mostly used in applications where reliability is critical, such as in space vehicles, submarines, and in remote locations.

The basic component of a fuel cell is the catalyst that helps convert chemical energy into electrical energy. The most common fuel cell uses hydrogen and oxygen with platinum serving as the catalyst. There has been tremendous research into developing better catalysts to increase the life, reliability, and efficiency of fuel cells. The most common type of fuel cell is the so-called proton exchange membrane fuel cell, though carbon nanotubes are also now being used as catalysts.

The applications of fuel cells include the standard applications in spacecraft (where hydrogen and oxygen, stored in fuel tanks, are easily available) and submarines, as well as in electrical vehicles and even experimental aircraft! Electrical or hybrid vehicles are already a reality though they still cannot compete with the conventional internal combustion engine.

## SURFACE CHEMISTRY

The work of German scientist Gerhard Ertl (b. 1936) has been crucial in not only understanding the working of fuel cells, but also in the development of more efficient cells. Ertl's work focuses on a major area of chemistry called surface chemistry. Surface chemistry attempts to study and understand how surface processes (like the rusting of iron, working of catalytic converters, etc.) work. It finds applications in the chemical industry (since many surface reactions work with catalysts), the semiconductor industry, as well as in the understanding of phenomena such as ozone depletion in the atmosphere. Ertl received the Nobel Prize in 2007.

Ertl is also responsible for developing the methodology for the study of surface reactions in detail. Using techniques from

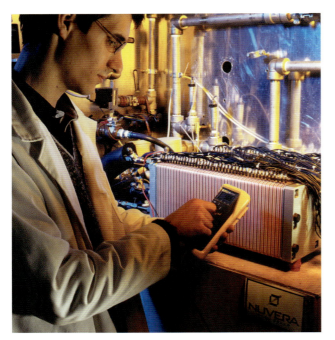

**FUEL CELL RESEARCH**

Technician studying the performance of a hydrogen fuel cell. This fuel cell combines hydrogen and oxygen (from air) to create electricity, heat and water as a by-product. Unlike batteries, fuel cells require no recharging.

fiscal incentives being provided by governments to renewable sources of energy.

## FUEL CELLS

Among the most promising sources of non-polluting energy is the fuel cell technology, which promises to deliver electrical power using hydrogen and oxygen with only water as a waste product. Fuel cells are basically devices that use up some reactants and convert chemical energy into electric power. They have been known since the middle of the 19th century though serious research on them only began in the 1950s. This was partly due to the demands of the space program where electric power is needed to run essential

several fields like semiconductor production, he was able to develop several different experimental procedures that together can provide a complete understanding of the surface reactions. This has proved to be important for developing better catalysts for more efficient fuel cells.

A major issue confronting the integration of fuel cells into the mainstream is the availability of hydrogen. Hydrogen required for the commonly used fuel cells needs to be made from hydrocarbons (typically from methane) and this process itself generates greenhouse emissions. There has been some progress in methods of making on-board hydrogen that does away with emissions and also providing for an elaborate hydrogen storage infrastructure. For instance, there have been reports of production of hydrogen sufficient for the on-board fuel cell by reacting sodium borohydride with borax. It is expected that the issues of reliability and efficiency as well as the manufacture of hydrogen will be resolved in the next few years. With the development of better catalysts, fuel cells could very well replace the internal combustion engine to some extent, thereby decreasing the emissions from, at least, the transportation sector.

## NANOTECHNOLOGY

The development of instruments like the atomic force microscope in the last decades of the 20th century gave rise to a whole new field of nanotechnology. Nanotechnology is the study and manipulation of matter at the nanometer (a billionth of a meter) scale. Although there have been few popular applications of nanotechnology, it is widely recognized as an area that will potentially revolutionize fields as diverse as medicine, manufacturing, and construction.

Chemical techniques, in an area known as synthetic chemistry, are already capable of synthesizing small molecules of a variety of compounds. These are being used in the chemical industry and the pharmaceutical industry. The challenge now is to assemble these simple individual molecules into larger structures or assemblies of molecules that have the desired structure and property. Such an assembling is possible because many molecules have an automatic affinity to certain other molecules. This is most dramatically seen in how the base pairs in DNA pair up, or even in how proteins fold in certain ways due to the properties of their component amino acids.

**GERHARD ERTL AT THE FRITZ-HABER INSTITUTE IN BERLIN**

German scientist Gerhard Ertl talks to the media after he won the 2007 Nobel Prize in chemistry for studies of chemical processes on solid surfaces, research that has advanced the understanding of why the ozone layer is thinning.

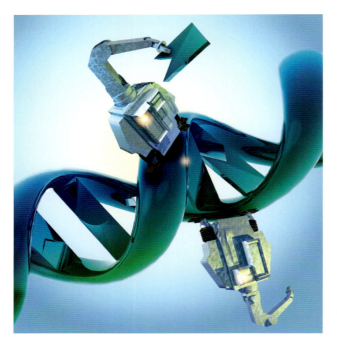

**DNA REPAIR USING NANOBOTS**

Computer artwork depicting the possibility of using robots at an atomic scale (nanobots) to repair damaged DNA. Surgery using infrared lasers and nanomaterials is already being attempted. Implants, especially neurological implants, made of nanomaterials are another possibility.

A chemical process known as sol-gel process is also now being widely used to produce nanoparticles. The technique has been known for over a century but has only recently been used extensively for producing nanoparticles. The basic idea is to start with a solution which then forms an amorphous gel under certain conditions. Typically, metallic compounds like metal chlorides or meal alkoxides are used as the solution, which, after going through certain reactions, form a colloidal suspension (a mixture where one substance is dispersed evenly throughout another). This suspension can be further processed to get the desired particles.

This process is used extensively in ceramics as well as in producing thin films of matter deposited on substrates that are then used for fabrication of electronic chips, among other uses. Materials made by this process also find applications in medicine, where these can be used for timed and even specific-location drug delivery, though it is not clear whether all the long-term safety issues have been addressed.

Another kind of nanoparticle is what is termed the quantum dot. These are essentially semiconductors that have properties different from the usual semiconductors. They are fabricated using various chemical processes and find applications in a variety of fields. In electronics, they are used for making diodes, transistors, more efficient solar photovoltaic cells, and even light-emitting diodes. In medicine, they are being used for the imaging of biological tissue inside the human body. There are, once again, unresolved issues of safety of these quantum dots for use in the human body, but with better technology, these will hopefully be resolved.

Nanoparticles, or particles of nanometer scale, have been part of the ceramics and metallurgy industries for several centuries. However, no systematic study, characterization, and production was possible until recently. Typically, two kinds of processes are used to make nanoparticles—a physical process called attrition, and a chemical process called pyrolysis. In attrition, larger particles are physically broken down to smaller particles using special techniques. In pyrolysis, the substance is heated and the waste products collected. Typically, a liquid or a gas of the material is forced at high pressure through a narrow opening and burned. The "soot" is collected and can be used to extract the required particles.

Nowadays, another technique is used to make nanoparticles. This is the thermal plasma technique in which a solid powder is heated to very high temperatures and on cooling, the nanoparticles are formed. Heating can be done either in an arc furnace or even by using radiofrequencies.

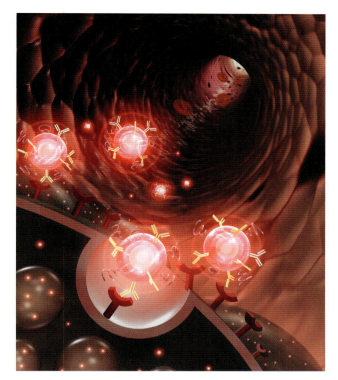

**QUANTUM DOT**

Quantum dot nanoparticle probes are used to target and image tumors that cause cancer. In this image, several quantum dots are attaching themselves to a tumor on the wall of a blood vessel. Quantum dots are also used as semiconductors and are widely used in manufacturing processes in the electronics industry.

The medical industry sees a great deal of potential in the applications of nanotechnology to diagnose, treat and research diseases. Improved, targeted, and timed drug delivery using nanoparticle probes and incision tools is being studied extensively. Similarly, imaging techniques, especially in oncology, are being explored using quantum dots. Micro, or more appropriately nano, surgery is also being attempted using infrared lasers together with nanomaterials. Implants made of nanomaterials are likely to be used very soon. More futuristic applications include nanorobots and other nanoscale machines for diagnostic and clinical purposes.

## GREEN ENERGY

The enormous progress made in chemistry in the past century has also raised concerns about the production of hazardous wastes and by-products. From carbon dioxide, which contributes to the greenhouse effect, to CFCs (chlorofluorocarbons) that are partially responsible for ozone depletion in the atmosphere, chemical wastes have a significant impact on human life. Huge chemical complexes routinely produce significant amounts of wastes that ultimately contaminate the air, soil, and water, and pose grave threats to the ecosystem.

Green chemistry is a relatively recent field of study within chemistry and aims to eliminate or at least reduce the amount of hazardous wastes in chemical processes and industries. The phasing out of CFCs after the Montreal Protocol (1987) led to the development of several non-hazardous refrigerants that have been successfully integrated commercially. CFCs were also used in producing polystyrene foam, used for packaging, and these have now been replaced by a process in which carbon dioxide can be used.

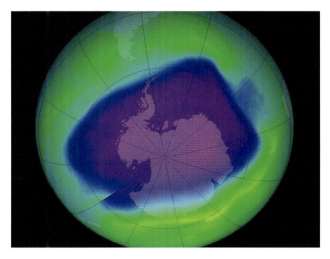

**ANTARCTIC OZONE HOLE**

This 2006 satellite image shows the low ozone levels over Antarctica. The color scheme progresses from purple (lowest) to yellow (highest).

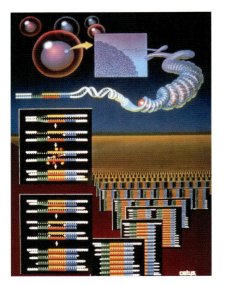

**POLYMERASE CHAIN REACTION**

Ilustration showing how the polymerase chain reaction (PCR) can amplify a sample of DNA to produce millions of copies. The two strands of the sample DNA are split apart (right box) and DNA primers (blue and green) added. The area to be amplified (orange and yellow) lies between the primers, and is filled in with the chemicals in DNA using the enzyme polymerase.

the development of techniques like PCR (polymarese chain reaction), genetic engineering became a reality.

Biotechnology finds applications in a variety of areas like medicine and agriculture. In medicine, biotechnology has come a long way since the company Genentech (founded in 1976) produced human insulin using genetic engineering techniques. This was done by inserting the insulin producing gene into the genetic material of the well known bacteria, E coli. Since E coli reproduces very rapidly, a large number of bacterial offspring of the genetically engineered mother soon emerge, all of which produce insulin.

In recent years, similar methods have been used to produce a variety of therapies for diseases like hepatitis, cancer, and osteoporosis. In addition, biotechnology has been used to make diagnostic aids for specific diseases. These are based on specific-molecule analysis techniques.

Another example of green chemistry is the introduction of a process of making a usable polymer from lactic acid, which can be easliy made from fermenting corn. Making polymers in this fashion avoids the use of fossil fuels that are typically used in the manufacture of polymers. Several such new processes are replacing existing industrial processes that produce large amounts of hazardous waste. In future, as our awareness of the hazards of many of the wastes from the chemical industry grows, we will be seeing more and more developments in green chemistry that will make the chemical industry not only more efficient, but also less hazardous.

## BIOTECHNOLOGY

The last couple of decades of the 20th century saw the emergence of biotechnology as one of the major areas of research and development. Although human beings have always used biotechnology to some degree ever since they started practicing agriculture, it was only in the mid 20th century that the deciphering of the genetic code provided a sound theoretical basis for biotechnology. Subsequently, with

**GENETICALLY MODIFIED ANTIBODIES**

A 150-litre research fermentor containing a strain of Escherichia coli, or E coli, bacteria. E coli strains are genetically modified to make antibodies, proteins that are part of the body's immune system.

## GENETIC THERAPY

A growing area in genetic medicine is that of genetic testing. The idea behind this is to determine the complete sequencing of a patient's DNA to identify the genes responsible for specific diseases and ailments. The genes responsible for the disease are usually mutated genes and these need to be identified. For this, two methods are used: in one, DNA probes that have the potential of attaching themselves to the pieces of mutated gene are used. If the mutated gene is present in the target DNA, then these probes will attach themselves to the sequence (since they are complimentary to those bases) and they can then be identified. In the second method, the sequence is compared with another healthy person's DNA to locate the mutated genes.

Genetic tests have been developed for a variety of diseases like Huntington's disease, certain forms of tumors, and also sickle cell anemia. Apart from the legal issues involved in genetic testing,

**GENE THERAPY RESEARCH ON MICE**

These mice are the site of experiments to find a cure for Duchenne Muscular Dystrophy (DMD), a condition typified by muscle wasting and loss of function. The mice had an abnormal gene that caused the muscles in their legs to waste.

there are also ethical issues involved that societies will have to resolve as the testing procedures become more accurate and are available for many more diseases.

An emerging area related to genetic testing is gene therapy. This is being used to treat a number of genetic and other diseases. The idea is to replace the defective genes responsible for the disease. This is done by either introducing the normal gene in non-reproductive cells or in the egg and sperm cells. This is done either by producing the genes outside the patient's body (called ex vivo) or producing them inside the body (in vivo). In the ex vivo method, cells from some part of the patient's body (typically blood or bone marrow) are allowed to be attacked by viruses that have the desired gene. The DNA of the virus gets integrated with the cell's DNA and these cells grow together. The cells are then injected back into the patient's body. In the in vivo method, the virus with the desired gene is inserted directly into the patient with the same effect.

As of now, gene therapy is not widely used in actual patients because of safety, ethical, and legal issues. However, many experiments are being carried out on animals and it is possible that in the next few years, some of the safety issues will be resolved.

## TRANSFORMING AGRICULTURE

The use of biotechnology in agriculture has progressed even more than in medicine. An area where biotechnology can potentially have an immense impact is the artificial increase in plant yields. This is crucial because even as the human and animal population is increasing rapidly, the arable area available for growing food is limited. Any yield growth, especially in crops like rice, wheat, and corn, can be greatly beneficial. However, there are technical problems that have still not been overcome. The yield of any crop is determined not by a single gene but by a large number of genes. Current techniques are able to transfer only a few genes into the crop, which is usually not enough to create a substantial increase in yield.

**GENETICALLY MODIFIED COTTON**

A crop of genetically modified (GM) upland cotton (*Gossypium hirsutum*). This cotton is a variety of the well-known Bt cotton, which was introduced in the US in 1996. The cotton is more resistant to pests.

## GENETICALLY MODIFIED CROPS

The improved productivity of genetically modified (GM) crops has led to a boom in their cultivation, especially in GM varieties of crops like corn, soybean, and cotton. The agrochemical multinational, Monsanto, has developed a strain of soybean (by transferring a gene from a bacteria into the genetic material of the soybean plant) that is resistant to herbicides. Similar examples exist in cotton, corn, and sugarcane. The spread of these GM varieties has increased enormously in the past few years. For instance, more than 85 per cent of the corn grown in the US is genetically modified.

Apart from making crops pesticide resistant, biotechnology has introduced several GM varieties that actually reduce the use of pesticides and herbicides required by the crop. This is done by the introduction of specific genes that produce chemicals toxic to certain common pests. The most well known example of this is Bt cotton, which has the gene from *Bacillus thuringiensis* (a bacteria found in the soil). This gene is responsible for the production of the Bt toxin, which is harmless to the plant but fatal for the insect. Several such examples now exist and it is likely that many of the common pests that attack commercial crops will be controlled in this manner in the future.

Genetical modification can also improve the resistance of crops to environmental stress. The majority of the world's agriculture (in terms of acreage) is still rain-fed and, hence, droughts are a major threat to food security. Genetically modified varieties could be made more resistant to drought and some research is already taking place to accomplish this. For instance, scientists have found that a particular gene, from a weed called thale cress, when inserted in tomato cells makes them more resistant to environmental stresses like drought or frost. However, it is still a long way before the major crops can be genetically modified to benefit large populations.

Agricultural production is crucially dependent on the protection of crops and plants from pests, which can destroy complete harvests over large areas. For most of history, humans had to live with this scourge, until the development of pesticides in the 20th century provided a way to defeat the pests. The use of pesticides led to a substantial increase in agricultural productivity and fuelled economic growth.

However, pesticide use in some cases can also be harmful to the plant. For instance, the use of pesticides that kill a certain pest in the maize plant, also tends to kill the plant itself. In the past couple of decades, biotechnology has been able to genetically modify many plants to make them more resistant to herbicides and pesticides. This means that pesticides can be used in quantities sufficient to eliminate the pests but without causing any harm to the plant.

Biotechnology can also make a difference in increasing the nutritional value of plants, improving their shelf life, appearance, and texture. A significant example of this is the recent development of Golden Rice. If rice, the staple food of a large part of the world's population, can be genetically modified in a way that some essential nutrients are added, it could be a major step forward for nutritional science. Scientists have been able to genetically modify a variety of rice to produce Golden Rice, which biosynthesizes beta carotene, a chemical that can be naturally converted to Vitamin A, an essential vitamin for the body. Improving the shelf life of fruits is another area where substantial progress has been made. This will be particularly beneficial to poorer countries where the wastage of crops due to poor transport and storage infrastructure is a major problem.

A much more challenging task for biotechnology is the use of plants for producing chemicals that have hitherto been created either synthetically or by using animals. Edible vaccines is an area witnessing a great deal of research. These vaccines can be administered as part of food, thus obviating the need for a sophisticated, and risky (in the case of syringes) delivery mechanism. Plants have been used to produce insulin at a substantially reduced cost than the conventional methods.

The unprecedented increase in oil prices in recent years has led to a renewed effort to produce fuel from biological sources. Biodiesel made from certain plants is already being produced even though its commercial viability is not certain. More significant is the production of ethanol, which

**CORN ETHANOL PROCESSING PLANT**

Piles of used corn grain at an ethanol production plant in Iowa, USA. Corn ethanol is a biofuel that is primarily used in the United States as an alternative to gasoline and petroleum. The use of biofuels is expanding in Europe and the Americas, especially in Brazil.

## ROGER KORNBERG

The transcription of the DNA onto the mRNA is done by a protein, an enzyme called RNA polymerase II, along with other enzymes. RNA Polymerase II and its associated proteins are critical elements of the cell machinery without which the transcription of genetic information and the synthesis of proteins is not possible. Roger D. Kornberg (b. 1947) and his collaborators at Stanford University used yeast to identify the process by which this transcription takes place. They were able to isolate the many proteins involved in the transcription process and thus understand the role of these proteins in the process. One of the major discoveries by Kornberg was that the process and its components are almost identical in all eukaryotes, i.e., organisms with a cell nucleus.

Not only did Kornberg elucidate the exact mechanism of transcription, he was also responsible for developing techniques for determining and visualizing the structure of complex proteins. He was able to see the three-dimensional structure of the RNA polymerase II using X-ray diffraction. In fact, the RNA polymerase II is the most complex known protein in terms of structure. Kornberg was awarded the Nobel Prize for chemistry in 2006 for this pioneering work. Interestingly, his father had won the Nobel Prize in 1959 for studying the mechanisms for the biosynthesis of DNA.

**AT THE 2006 NOBEL PRIZE CEREMONY**

Kornberg and his collaborators at Stanford University used yeast to identify the process by which the transcription of the DNA is done onto the messenger RNA (mRNA).

is normally derived from crops such as sugarcane and, in recent years, from maize. However, the use of sugarcane and maize for ethanol production has meant a diversion of these crops from food towards fuel and, hence, a huge increase in their price. A more sustainable alternative would be the development of bacteria that can be used to break down the cellulose in traditional waste products like baggasse to produce ethanol.

The use of biotechnology in agriculture and food production has also raised many ethical and ecological issues. Though the increase in yields of GM crops cannot be doubted, there are still unresolved issues about their safety. A large number of countries still do not permit the production or sale of GM foods. There have also been concerns about the ecological safety of GM crops, since there is a danger of new pathogens or genes being released into the wild. In a wider sense, there are also risks of GM varieties proving less hardy in the long run than natural ones, which have evolved over millennia and survived the tough tests of natural selection.

## BIOCHEMISTRY

Although biotechnology has tended to dominate the news in recent years, there have also been significant developments in understanding various cellular mechanisms and structures. One of the major advances in recent years has been an understanding of the transcription process by which genetic information in the DNA (in the nucleus) is translated onto the RNA. The transcription is actually done onto a form of RNA called the mRNA (messenger RNA), which then moves out of the nucleus to the ribosome, the site of protein synthesis in the cell. Proteins are made up of amino acids and what distinguishes one protein from another is the sequence of amino acids and the three-dimensional structure of the molecule.

## PHARMACEUTICAL BREAKTHROUGHS

An interesting fact about nature is that even though most biological molecules come in two forms—left-handed and right-handed, only one of the forms is used in biological processes. The two forms are called enantiomers, and while they may have the same chemical composition, their properties can be vastly different. For instance, the drug thalidomide, used for preventing nausea during pregnancy, has an enantiomer that can cause severe fetal damage.

It is observed that all the essential molecules for life, like amino acids, fats, proteins, and even the nucleic acids, are always found in one kind of enantiomer. This is strange,

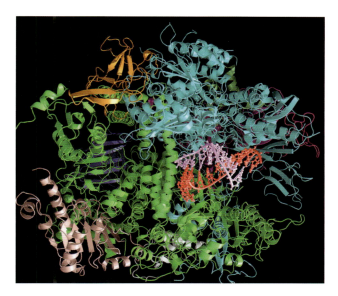

**DNA TRANSCRIPTION, MOLECULAR MODEL**

Secondary structure of the enzyme RNA polymerase II, synthesizing a strand of mRNA (messenger ribonucleic acid) from a DNA template.

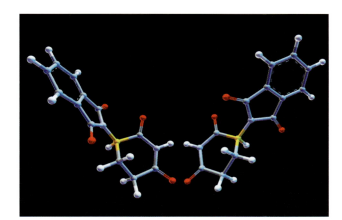

**LEFT- AND RIGHT-HANDED THALIDOMIDE**

Thalidomide is an enantiomer, a molecular structure with two sets of atoms, distinguished as left-handed and right-handed, that are mirror images of each other but with dramatically different chemical properties. Thalidomide, used as a sedative for pregnant women during the 1950s, was found to cause deformities in fetuses. Later research discovered that only the left-handed thalidomide caused the deformities.

since in the laboratory usually both the kinds of compounds are created, the so-called symmetric synthesis. Nature's preference for one-handedness might be a biological or evolutionary accident but it has profound implications that are yet to be fully understood. Nevertheless, the production of the "correct" enantiomers is a very important task for the pharmaceutical and food industry.

The work of W. Knowles, K.B. Sharpless, and R. Noyori in this area has been significant. They have been able to create catalysts that allow only one form of the compound to be created in certain kinds of reactions. This work is very important for synthesis in a variety of industries like pharmaceuticals, flavorings, and industrial chemicals. It has also had a seminal impact on the development of advanced materials. The three scientists shared the Nobel Prize for chemistry in 2001.

## ISRAELI SCIENTISTS HERSKO AND CIECHANOVE

Israeli scientists Avram Hersko (left) and Aaron Ciechanove, winners of the 2004 Nobel Chemistry Prize, speak to the media. The scientists, along with US scientist Irwin Rose, won the prize for their work on protein breakdown.

The interface of chemistry and biology has seen significant developments in the past few years. These developments have not only advanced the knowledge of fundamental life processes, they also hold the promise that drugs for some of the most dreaded and endemic diseases can be developed soon. For instance, I. Rose, A. Ciechanover, and A. Hershko have discovered the process through which a certain class of proteins called ubiquitin proteins are regulated. Ubiquitin is responsible for protein degradation, a process by which the cell discards unwanted proteins. The ubiquitin attaches itself to the protein that needs to be discarded, and then transports it to a place where the protein is divided into its amino acids. The interesting thing is that the ubiquitin molecule later detaches itself and can be reused. When this mechanism fails, diseases like Alzheimer's and Parkinson's develop. An understanding of the regulation of ubiquitin proteins can also lead to the development of better drugs for other diseases like cancer and cystic fibrosis. The three scientists were awarded the Nobel Prize in 2004.

## 2005 NOBEL PRIZE LAUREATES

(from left) Roy J. Glauber of Harvard University, Richard R. Schrock of the Massachusetts Institute of Technology, Robert H. Grubbs of the California Institute of Technology, Thomas C. Schelling of the University of Maryland, and John L. Hall of the University of Colorado, at their reception in Washington.

The development of catalysts for the metathesis method of organic synthesis is another significant step in producing new drugs. Developed by Y. Chauvin, Robert Grubbs, and R. Schrok, the catalysts are easy to fabricate and find use in the synthesis of many products, including specialized polymers. The development of these new catalysts has meant that processes in the chemical industry can be made more efficient. Metathesis applications now involve fewer reaction steps, fewer resources, and less wastage. The methods are also simpler to use because they are stable in air and at normal temperatures and pressures. They are also more environmentally friendly, as they use non-injurious solvents and generate less hazardous waste. This is an important landmark in "green chemistry". Chauvin, Grubbs and Schrok won the Nobel Prize for chemistry for this important work in 2005.

## TOWARDS A SUSTAINABLE CHEMISTRY

With all these technological innovations, the 21st century promises to be an exciting one. Green chemistry can reasonably expect to see the development of better processes that are not as wasteful of resources and do not produce as much hazardous waste as the existing procedures. Renewable energy sources, such as solar cells, will become increasingly popular and cheaper as a result of more advanced materials. New materials with unusual properties are likely to be developed and these could have applications in diverse areas.

The field of genomics and biotechnology is likely to see tremendous progress in the coming years. With the decoding of the human genome and the development of new techniques, whole new areas have opened up for the use of genomic applications. DNA sequencing will become even quicker and more accurate, leading to its widespread use. Inexpensive DNA chips will be available for quick analyses of genetic material. More and more genes responsible for diseases may be identified and therapies developed. In agriculture, biotechnology has the potential of revolutionizing the production of food in a way that could be as significant as the Neolithic revolution when agriculture first emerged.

**DNA SEQUENCING**

A petri dish containing banded DNA sequences or genetic fingerprints. The pattern of bands represents the chemical sequence that forms the genetic code for a section of DNA. The DNA sample is cut into fragments by an enzyme and the fragments placed into a gel in the petri dish and separated by the application of an electric charge. DNA sequencing is used in forensic science, medical research, and genealogy.

As we saw in the first chapter, the science of chemistry evolved from alchemy, which was obsessed with finding ways to change base metals into gold. Though this goal was in principle realized in the 20th century with the transmutation of elements, several other goals of chemistry remain unrealized. One hopes that the 21st century will see the use of chemistry to not only improve the quality of human life, but also make it more sustainable. No other era in human history has seen a greater danger of human beings destroying the very environment responsible for their existence.

## ELECTROSPRAY IONISATION

An electrospray ionisation chamber. The technique involves injecting a solution into a strong electric field, which disperses it into a fine spray of charged droplets. These droplets can be further broken down into free ions that a conventional spectrometer can analyze.

## JOHN B. FENN (B. 1917)

Fenn with his 2002 Nobel Prize for Chemistry at the award ceremony in Stockholm.

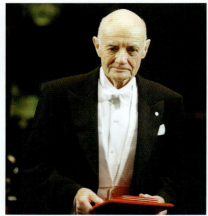

# JOHN FENN

The tremendous advances in chemistry would not have been possible without the development of new and innovative techniques that allowed researchers to identify and analyze compounds. One such technique is the electrospray ionization technique developed by the American analytical chemist John Bennett Fenn (b. 1917). The technique is a powerful extension of the conventional mass spectroscopy technique to identify molecules. Mass spectroscopy works by taking a sample in ionized gaseous form that can then be moved in a magnetic field to determine the mass (and hence the composition) of individual components. The limitation of mass spectroscopy is that it cannot be used for substances that are destroyed during the process of conversion to a gaseous form. These include many important compounds like proteins.

Fenn developed a way to convert samples of large molecules into gaseous form without such degradation. He invented electrospray ionization, a technique that involves injecting a solution of the sample into a strong electric field, which disperses it into a fine spray of charged droplets. As each droplet shrinks by evaporation, the electric field on its surface becomes intense enough to break individual molecules from the droplet, forming free ions ready for analysis using a conventional mass spectrometer. The ionization technique has proved to be a highly versatile technique, and has been used in the development of pharmaceuticals and other areas. In the pharmaceutical industry, it has led to an increased speed of identifying complex molecules, a potentially revolutionary step in the development of drugs for a variety of diseases.

# BIOLOGY

* ANCIENT BIOLOGY

* MEDIEVAL TO EARLY MODERN

* MODERN BIOLOGY

# Chapter 1
# ANCIENT BIOLOGY

**MEDICAL HIEROGLYPHS IN EGYPT**

The Egyptians had advanced medical practices, ranging from embalming to faith healing, to surgery and autopsy. They were the first to record their healing practices in the form of hieroglyphs and papyri. The papyri, containing magical formulae and remedies, are among the most important medical documents surviving from ancient Egypt.

**FACING PAGE:** A re-creation of how the ancient Egyptian medical manuscripts may have been used and created.

**B**iology is the study of living things and their relationship with each other and the environment. Subdivided into botany (the study of plants), zoology (the study of animals), morphology (the study of the structure of organisms), and physiology (the study of the functioning of living systems), biology is essentially the study of common biological processes in life forms. Given the evidence of domestication of plants and animals from very early times, it is clear that biology is a fairly ancient science.

## BIOLOGY IN MESOPOTAMIA AND EGYPT

Long before the Greeks, the ancient civilizations of Mesopotamia and Egypt, dating back to c. 3500 BC, studied nature and attempted to gain knowledge about natural phenomena. The Assyrian and Babylonian carvings and reliefs show aspects of veterinary medicine. The oldest Babylonian text on medicine is recorded in the Old Babylonian period in the first half of the second millennium BC. The most extensive Babylonaian medical text is the *Diagnostic Handbook*, written by Esagi-kin-apli of Borsippa, a physician during the reign of Adad-apla-iddina (1069–1046 BC). The Babylonians had learned that the date palm reproduces sexually and that pollen from the male plant would fertilize female plants. A Babylonian record of the Hammurabi period (c. 1800 BC) mentions the male flower of the date palm as an article of commerce, and descriptions of date harvesting go back to about 3500 BC.

The Egyptian civilization, starting in c. 3300 BC and lasting till the Persian invasion in 525 BC, was the pioneer of medicine. Papyri and artifacts found in the tombs and pyramids of the Egyptians indicate that they possessed considerable medical knowledge. The well-preserved mummies testify to their knowledge about the human anatomy and the preservative properties of plants used for embalming. Plant necklaces and bas-reliefs reveal that the Egyptians were aware of the medicinal qualities of plants and used them n their healing practices. Besides medicinal purposes, the Egyptians put plants to other uses too. For instance, they produced paper out of the pith of papyrus (*Cyperus papyrus*).

Egyptian physicians were aware of the importance of the pulse, and of a connection between the pulse and the heart. The earliest known surgery in Egypt was performed around 2750 BC. Imhotep of the Third Dynasty (*fl.* 27th century BC), called the founder of ancient Egyptian medicine, is sometimes credited with being the original author of the *Edwin Smith*

**THE EBERS PAPYRUS,** Library, University of Leipzig, Germany

Compiled in Egypt in about 1570 BC, *The Ebers Papyrus* is the oldest preserved medical document, containing the most complete description of the ancient Egyptian medicine known. The scroll contains 700 formulae and traditional remedies to cure afflictions ranging from crocodile bite to toenail pain. It also includes a fairly accurate description of the circulatory system, mentioning the existence of blood vessels throughout the body and the heart's function as the center of the blood supply. There are chapters on medical specialities such as dentistry, intestinal disease, eye and skin disorders, and surgery. Many specific disorders like diabetes and arthritis are described. The papyrus was found in Thebes in 1862 by the German George Maurice Ebers. The other significant medical work from ancient Egypt is *The Edwin Smith Papyrus*, a book of surgery by Imhotep, acquired by Edwin Smith, an American, in 1862.

*Papyrus*, the world's earliest surviving medical text, detailing cures, ailments, and anatomical observations, including an oblique reference to the cardiac system. *The Ebers Papyrus* (c. 1500 BC), a compilation of ancient Egyptian remedies, provides a fairly accurate description of the circulatory system and the evidence of awareness about tumors. Hesyre of Egypt, "Chief of Dentists and Physicians" for Pharaoh Djoser of the Third Dynasty in the 27th century BC, is one of the earliest known physicians in the world.

**3200 BC** Civilization begins at Sumer.

**3100–2686 BC** Early Egyptian dynastic period; unification of Upper and Lower Egypts.

**2900–2000 BC** Bronze Age. Beginning of early Aegean cultures.

**2900 BC** First Egyptian hieroglyphs.

**2500 BC** Minoan civilization develops.

**1050–750 BC** Early period of Greece—Greeks send settlers to create colonies in Asia Minor.

**753 BC** The founding of Rome.

**612 BC** Nineveh (the capital of Babylonia) is captured, marking the end of the Assyrian empire.

**499 BC** Greek city-states revolt against the Persian rule.

**492–449 BC** The Persian Wars.

**450 BC** Sushruta writes the *Sushruta Samhita*.

**450 BC** Xenophanes examines fossils and speculates on the evolution of life.

**400 BC** Hippocrates establishes the school of medicine at Cos.

**350 BC** Aristotle attempts a comprehensive classification of animals.

**332 BC** Alexander of Macedonia defeats Persians and builds a capital at Alexandria.

**300 BC** Theophrastus begins the systematic study of botany.

**300 BC** Herophilus dissects the human body.

**300 BC** Alcmaeon of Croton distinguishes veins from arteries and discovers the optic nerve.

**250 BC** Erasistratus of Loulis dissects the brain and distinguishes between the cerebrum and the cerebellum.

**100 BC** Diocles coins the term anatomy.

**AD 77** Pedanius Dioscorides completes writing *De Materia Medica*.

**AD 77** *Historia Naturalis* by Pliny the Elder is published in 37 volumes.

**130–200** Galen writes numerous treatises on human anatomy.

**324** Constantine the Great establishes his capital at Byzantium (Constantinople).

**332–395** Graeco-Roman period.

**455** Vandals sack Rome.

**476** Western Roman Empire ends.

## CHINESE MEDICINE

The Chinese emperor Shen Nung (*fl.* 28th century BC) is believed to have written *Divine Husbandman's Materia Medica* in *c.* 2800 BC, listing 365 medicines that could be derived from minerals, plants, and animals, and including descriptions of many important food plants, such as the soybean. The ancient Chinese produced silk from the silkworm *Bombyx mori* and understood the principle of biological pest control, employing certain ants to destroy insects that bored into trees.

China also developed traditional medicinal practices. Much of the philosophy of traditional Chinese medicine was derived from empirical observation of disease and illness by Taoist physicians. It reflects the classical Chinese belief that individual human experiences express causative principles found in the environment at all levels. During the Han dynasty, Zhang Zhongjing (*c.* 150–219), who was the mayor of Changsha near the end of the 2nd century AD, wrote the *Treatise on Typhoid Fever.* Another prominent physician of this period was Hua Tuo (*c.* 140–208), who anesthetized patients during surgery with a formula of wine and powdered hemp. During the 3rd century AD, the Jin Dynasty practiced and advocated the acupuncture technique.

## BOTANY IN ANCIENT INDIA

The Indus Valley people (*c.* 3000–1500 BC) cultivated crops, including wheat, barley, millet, dates, vegetables, fruits, and cotton. They worshipped trees and used plant extracts for coloring and glazing. They used manure and practiced the rotation of crops for improving the fertility of soil and providing nourishment to plants. Relics of the ancient civilization of Mohenjodaro and Harappa, dating to 2500 BC and earlier, show that the people of north India had a well-developed science of agriculture. The medicinal properties of plants were also known. A document dated about the 6th century BC describes the use of about 960 medicinal plants and includes information on such topics as anatomy, physiology, pathology, and obstetrics.

**HUA TUO,** Artwork from *Research on Chinese Superstitions* by Henri Dore, Shanghai, Bibliotheque Nationale de France

Hua Tuo was a well-known Chinese surgeon during the Han dynasty. He invented an oral anesthetic that made the patients unconscious, allowing him to operate them to remove tumors and even perform internal abdominal surgery. Similar anesthetics would be used in Western medicine only in the 19th century.

Medicine, agriculture, and horticulture developed to a great extent during the Vedic period in India (*c.* 2500–600 BC). A classification of plants is attempted in the Vedic literature. The *Rigveda* mentions that Vedic Indians had some knowledge about the action of light on plants. In the post-Vedic period, there is evidence that botany developed as an independent science, subdivided into the science of medicine (as embodied in the *Charaka* and *Sushruta Samhita*), agriculture (as embodied in the *Krsi-Parasara*), and arbori-horticulture (as illustrated in the *Upavana-Vinoda*). *Vrksayurveda* by the sage Parasara is one of the oldest works in field of botany, written between the 1st century BC and the 1st century AD. It deals with irrigation, farm implements, land preparation, plant propagation, plant nutrients, plant protection, disease management, and harvesting and storage of crops.

The *Atharva Veda* is said to be the earliest recorded authority on plant morphology. It describes eight types of growth habits of trees. Though it was known that plants were made of the root, shoot, stem, twigs, flowers, and fruits, ancient Indian literature further classified roots on the basis of their growth behavior and structure, like, roots originating from the branches, multiple roots, plants with three main roots, thick root, thin root, and fasciculate root. Leaves were also described in great detail on the basis of their shapes, color, and texture.

In the *Rigveda*, *daru* or the hard wood is distinguished from the softer outer part of a tree. In the *Vrksayurveda*, Parasara describes tissue systems meant for the transportation of nutrients and sap. The vascular system is divided into two categories, which correspond to the xylem and phloem we know today. Parasara also offers a detailed description of the plant cell. According to him, the internal structure of the leaf consists of innumerable compartments, which are filled with sap. These are surrounded by a cell membrane (or cell wall).

Scholars in the Vedic period had knowledge that plants received their nutrients from the soil in the form of a solution through the root. *Santiparva* explains the phenomenon of the ascent of sap and states that water absorbed by the plant is converted into food under the influence of *agni* (energy) and *maruta* (air), and due to this, the plant can grow. The seed (*vija*) was known to be enclosed in a vessel called *vijakosa*, and the endosperm and the cotyledon were described. Having recognized monocotyledonous and dicotyledonous seeds, Parasara said that the seedling receives nourishment from the cotyledons during sprouting. This nourishment enables the seedling to grow until its root develops. The cotyledons dry up when the seedling is able to receive nourishment directly from the soil.

It was recognized that plants had male and female parts and that some species had separate male and female plants, as in *Pandanus odoratissimus* and *Hollerhina antidysenterica*. The *Atharva Veda* and *Arthasastra* describe propagation of plants by seeds, roots, cuttings, and grafting. Plants were classified according to their botanical characters—like the shape and color of leaves, flowers, and roots—and properties, especially medicinal. The *Rigveda* divided plants roughly into four broad

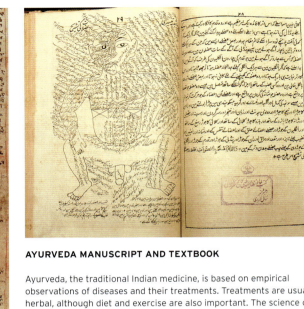

**AYURVEDA MANUSCRIPT AND TEXTBOOK**

Ayurveda, the traditional Indian medicine, is based on empirical observations of diseases and their treatments. Treatments are usually herbal, although diet and exercise are also important. The science of ayurveda developed in ancient India and is popular even today. The manuscript on the left, written in Sanskrit and illustrated with astrological symbols and mythological characters, is meant for the instruction of physicians in ayurvedic medicine.

**Top:** *The Canon of Medicine* by Avicenna (972–1036), published in 1593, is a textbook on ayurveda. It is still used by students at the Unani Medical College in Hyderabad, India.

classes: trees, herbs, creepers, and grasses. In his *Vrksayurveda*, Parasara classified plants into distinct families like Leguminosea (plants bearing legumes or pods) and Rutaceae (plants bearing spines, odoriferous leaves, and winged petioles).

## GREEK BIOLOGY

One of the earliest Greek philosophers, Thales of Miletus (*fl.* 6th century BC) postulated that the world and all living things were made from water. Anaximander (*c.* 610–545 BC), a student of Thales, proposed that life arose spontaneously in mud and that the first animals to emerge were fish with a spiny skin. According to him, these fish eventually left water and moved to dry land, where they gave rise to other animals by transmutation.

In southern Italy, an important school of natural philosophy was founded by Pythagoras (*c.* 580–500 BC) in *c.* 500 BC at Crotone. Alcmaeon (*fl.* 6th century BC), a student of Pythagoras, investigated animal structures and pioneered anatomical dissection. He studied the development of the embryo, described the difference between arteries and veins, discovered the optic nerve, and recognized the brain as the seat of the intellect.

The Greek physician Hippocrates (*c.* 460–377 BC), credited with rejecting the divine notions in the favor of rational medicine, established a school of medicine on the Mediterranean island of Cos around 400 BC. His physiology was based on the humoral theory, whereby the four humors or bodily fluids—blood, phlegm, yellow bile, and black bile—were required to be kept in balance. He recognized how the environment can influence human nature and suggested that extreme climates tend to produce a powerful type of inhabitant, while an even, temperate climate is conducive to indolence. *The Hippocratic Corpus*, the collection of medical works attributed to Hippocrates, contains

**ARISTOTLE,** Joos van Ghent, *c.* 1475

Aristotle is the first true biologist as he originated the scientific study of life. Though *The Hippocratic Corpus* deals with human anatomy, its focus is exclusively on human health. By contrast, Aristotle considered the investigation of living things, and especially animals, central to the study of nature. His zoological writings are the first systematic and comprehensive study of animals.

treatises by several authors on a variety of medical topics, including diagnosis, epidemics, obstetrics, pediatrics, nutrition, and surgery.

Around the middle of the 4th century BC, ancient Greek science reached a climax with Aristotle (384–322 BC), who was the most influential scholar of antiquity in all branches of knowledge, including biology. His treatises on human anatomy have been lost, but his major works on animals—*The Parts of Animals*, *The Natural History of Animals* and *The Reproduction of Animals*—list important anatomical observations. Aristotle attempted a system of animal classification. Using two broad categories—animals with blood and those without blood—he identified about 540 animal species. The animals with blood included those now grouped as mammals, birds, amphibians, reptiles, and fishes. The bloodless animals comprised the cephalopods, the higher crustaceans, the insects, and the testaceans (a group of all the lower animals). Aristotle was the first to show any understanding of an overall systematic taxonomy.

**PEDANIUS DIOSCORIDES,** Colored engraving from *Vies des Savants Illustres*, 1866

Dioscorides (at left and at right) is seen receiving the mandrake plant from the nymph Euresis. When working as a surgeon in the Roman army of Nero, Dioscorides used the mandrake plant, combined with wine, as an anesthetic when carrying out surgical procedures. He is best known for his five-volume medical treatise *De Materia Medica*, in which he described about 600 herbs and 1000 herbal remedies. The book served as the primary source of pharmacology and remained in use, in various translations, until 1600.

Aristotle's careful study of animals led to many lasting observations: the embryo in animals is the equivalent of the seed in plants; the sex of the embryo is determined at an early stage in its development; the fetus is fed through the umbilicus in mammals; mammals have lungs, breathe air, are warm-blooded, and suckle their young; some reef fish only exist as females, and the dominant female transmutes to male. He correctly described the biology of the honeybee, the embryology of the chick, and the anatomy of the hyena. He contributed to modern embryology by discussing the difference between epigenesis and preformation, and anticipating the theory of cell-streams. A large part of Aristotle's work was devoted to reproduction. He observed that males compete for the attention of females; female mammals eat more when they are pregnant; the sperm of the last male to mate fertilizes the egg among birds; and large animals are less fecund than small ones.

Aristotle contributed to the theory of evolution by postulating the principle of adaptation, stating that all organisms structurally and functionally adapt to their habitats. He identified structural homology and functional analogy in organs. For instance, organs like the hand, claw, and hoof are analogous structures. He observed that general structures appear before specialized ones and that tissues differentiate before organs.

Aristotle is believed to have written on botany as well but these works have not survived. However, the work of his successor Theophrastus (c. 372–287 BC), *De historia et causis plantarum*, has survived as the most important contribution of antiquity to botany. It describes the morphology, natural history, and therapeutic uses of plants. Theophrastus distinguished between the external parts (called organs) and the internal parts (called tissues). He attempted a scientific nomenclature, with some names having survived into modern times, such as *karpos* for fruit and *perikarpion* for seed vessel. Although he mentioned over 500 plants, Theophrastus did not propose an overall classification system for plants, but clustered species into groups, which could be considered precursors of today's genera. He wrote about sexual reproduction in flowering plants and recorded observations on seed germination and development.

## ROMAN MEDICINE

After the Greek period, new centers of learning came up. The most famous was the library and museum in Alexandria, where all the significant biological advances were made from c. 300 BC until around the time of Christ. Herophilus of Chalcedon (c. 335–280 BC) was one of the most outstanding physicians from Alexandria, who recognized the brain as the center of the nervous system and described it in detail. He established the diagnostic value of the pulse and started using a water clock to measure its frequency. Based on his knowledge, he wrote several books, including a general treatise on anatomy, a book on the eyes, and a handbook for midwives.

Unlike Herophilus, Erasistratus of Loulis (c. 325–250 BC), a younger contemporary, discarded the humoral theory of Hippocrates. He distinguished between pulmonary and systemic circulation and described the bicuspid and tricuspid valves. Although he was wrong about blood flowing from the veins into the arteries, he was correct in his conjecture that the body had small interconnecting vessels. He described the brain more accurately than Herophilus, differentiating the cerebrum from the cerebellum and sensory from motor nerves. Rejecting the theory that nerves are hollow and filled with pneuma (air), he was the first to state that nerves are solid, full of marrow.

Pedanius Dioscorides (c. AD 40–90) was a Greek physician, whose travels with the armies of Roman emperor Nero gave him the opportunity to study the medicinal qualities of many plants and minerals. He wrote *De Materia Medica* in AD 77, the classical source of botanical terminology and pharmacology.

Written in five books, the work, comprising information on about 1,000 natural drugs, was an authoritative text on herbal medicine till the end of the 15th century.

Around the same time as Dioscorides, Pliny the Elder (AD 23–79) completed *The Natural History*. Divided in 37 books, the work deals with different branches of biology: zoology, botany, and medicine. Although the work did not have a lasting influence, it was an authoritative text till the end of the 17th century because of the lack of more reliable information.

Perhaps the last of this school was Galen of Pergamum (c. 129–216), a Greek physician who practiced in Rome in the 2nd century AD. As the dissection of humans was not considered proper, Galen performed extensive dissections and vivisections on animals. He studied the muscles, the spinal cord, the heart, the urinary system, and proved that the arteries are full of blood. Galen's knowledge of human anatomy was faulty, but his work *On the Natural Faculties* was the authority on medicine for centuries to come.

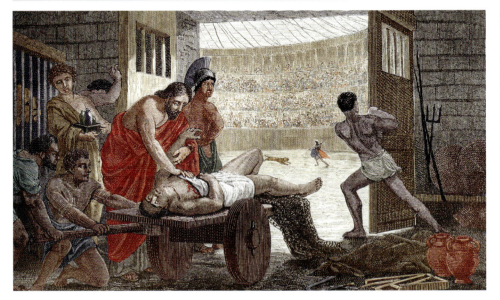

**GALEN TREATING A GLADIATOR IN PERGAMUM,** *Vives des Savants Illustres*, 1866

Galen, an ancient Greek physician, served as surgeon to the gladiators in Pergamum, his birthplace. He left Pergamum in AD 162 and emigrated to Rome, where he earned a brilliant reputation as a practitioner and a demonstrator of anatomy. He distinguished seven pairs of cranial nerves, described the valves of the heart, and observed the structural differences between arteries and veins. One of his most significant observations was that the arteries carry blood, and not air, as had been taught for 400 years. His ideas dominated western medicine for over a millennium.

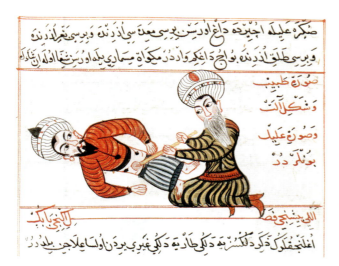

## ISLAMIC BIOLOGY

After the decline of the Roman Empire, science lay dormant in Europe for a thousand years and the Arabs became the custodians of science and dominated biology, as they did other disciplines. From the 3rd until the 11th century, biology was essentially an Arab science. Although Muslim scientists were not great innovators, they studied the works of Aristotle and Galen and translated them into Arabic. Medieval Islamic physicians, scientists, and philosophers made significant contributions to biological knowledge between the 8th and 13th centuries, during what is known as the Islamic Golden Age .

During this period, botany was favored over zoology and great attention was paid to horticulture, as evident in the beautiful gardens of this time. In zoology, the horse was the preferred animal of study. Abu Ubaidah (AD 728–825) devoted more than half of his 100 books to the study of the horse. In Basra, Al-Jahiz (AD 776–869), a noted zoologist, wrote the *Kitab al Hayawan* (*Book of Animals*), which explored concepts of evolution, adaptation, and animal psychology. He was

**EARLY SURGERY,** From an Ottoman translation of a surgical work, by Sharaf ad-Din, Turkish School, 1466

Surgical puncture of the abdominal cavity of the aspiration of peritoneal fluid with a canula on a patient suffering from dropsy.

**FACING PAGE:** Medicinal Plants, From *Traite de Medecine* by Claudius Galenus (c. 130–201), Translated by Gerard de Cremone (1114–1187), Bibliotheque Municipale, Laon, France

the first to note changes in bird life through migrations and described the method of obtaining "ammonia from animal fat by dry distilling." Al-Damiri (1344–1405) wrote the *Hayat Al Hayawan* (*Life of Animals*), which is an encyclopedia on animal life. Al-Masudi (888–957) made an attempt at the theory of evolution in his well known works *Meadows of Gold* and *Kitab al-Tanbih wal Ishraq*, where he suggested that evolution began from mineral to plant, from plant to animal, and from animal to man.

The Moors in Spain made significant contributions to botany. They discovered sexual difference in plants like palms and hemps, and were keen collectors of botanical specimens. They classified plants into those that grow from seeds, those that grow from cuttings, and those that grow of their own accord, that is, wild growth.

The Spanish Muslims augmented the herbology of the Greeks by the addition of 2,000 plants and laid out botanical gardens in Cordoba, Baghdad, Cairo, and Fez for teaching and research. For instance, Al-Ghafiqi of Cordoba (d. 1165) collected and described plants from Spain and Africa. Ibn Sina (c. 980–1037),

**800** Al-Jahiz introduces the idea of evolution, adaptation, and food chain.

**850** Al-Dinawari describes more than 600 plants in his *Book of Plants*.

**1010** Ibn Sina publishes *The Canon of Medicine*, introducing clinical trials and clinical pharmacology.

**1095** Christian Crusades to Jerusalem begin.

**1200** Abu al-Abbas al-Nabati develops an early scientific method for botany.

**1455** Gutenberg invents the working movable type.

**1517** Martin Luther pens 95 Theses, leading to the Reformation.

**1542** Leonard Fuchs compiles a herbal dictionary.

**1543** Andreas Vesalius publishes the anatomy treatise *De Humani Corporis Fabrica*.

**1597** John Gerard publishes *The Herball* or *General History of Plants*.

**1628** William Harvey demonstrates the circulation of blood in animals.

**1642** Civil war breaks out in England.

**1651** William Harvey states that all animals, including mammals, develop from sperm and eggs.

**1653** Nicholas Culpeper publishes *The English Physician Enlarged*.

**1658** Jan Swammerdam observes red blood cells under a microscope.

**1663** Robert Hooke sees cells in cork using a microscope.

**1669** Marcello Malpighi establishes the field of microscopic anatomy.

**1672** Malpighi publishes the first description of the development of a chick embryo, including the formation of muscle somites, circulation, and nervous system.

**1676–1683** Anton van Leeuwenhoek observes protozoa, spermatozoa, and bacteria. Leeuwenhoek's discoveries renew the question of spontaneous generation in microorganisms.

**1735** Carolus Linnaeus publishes *Systema Naturae*, his first work on botanical classification.

**1753** Linnaeus introduces the binomial system of nomenclature in botany.

**1809** Jean-Baptiste Lamarck proposes Lamarckism, a modern theory of evolution based on the inheritance of acquired characteristics.

**BOTANICAL PLANTS,** From *The Wonders of the Creation and the Curiosities of Existence* by Zakariya-ibn Muhammed al-Qazwini, 13th century, Institute of Oriental Studies, St Petersburg, Russia

also known as Avicenna, was a Persian scientist around the beginning of the 11th century. His writings on medicine and drugs were particularly authoritative and remained so until the Renaissance. He reintroduced the works of Aristotle to Europe, where they were translated into Latin from Arabic.

As agriculture developed in the Islamic world, cash cropping and crop rotation systems were introduced. Agriculture had a scientific underpinning, relying on highly developed irrigation techniques and a large repertoire of crop varieties. One of the early Arabic works on agronomy and agriculture was *Nabatean Agriculture* by Ibn-Wahshiyya (*fl.* 9th/10th century). In the 12th century, Ibn-al-Awwam Al-Ishbili (*fl.* 12th century) wrote the *Kitab al-Filaha*, which combined his knowledge of agriculture with that of the Nabatean Agriculture and his other Arabic predecessors. In this book, he described more than 585 plants, including 50 fruit trees, discussed plant diseases and their remedies, and made observations on the properties of soil and the types

**JOHN GERARD**

Gerard's *The Herball*, first published in 1597, was revised and enlarged by Thomas Johnson in 1633. Gerard used the *Materia Medica* of Dioscorides, the works of the German botanists Fuchs and Gesner, and the Italian Matthiolus (1501–1577). Divided into three books, this monumental work contains about 2850 plant descriptions and 2700 illustrations.

of manures they need. This work was later translated into Spanish by Banqueri in Madrid (in 1801) and into French by Clement-Mullet in Paris (in 1864).

## RENAISSANCE TO MODERN

During the European Renaissance, there was a growing interest in natural history and physiology, but botanical research was done primarily to support healing and medicine. The 16th century witnessed the spread of botanical knowledge. A significant feature of this period was the publication of a number of floras of different countries. The first European universities—Leiden (1577), Montpellier (1593), and Heidelberg (1597)—redirected attention from the study of plants through manuscripts to the examination of living plants. The central focus of research became the diverse floras and the foreign plants that were cultivated in botanical gardens. Luca Ghini (1490–1556) from Pisa is credited with being the first to press and dry plants in order to conserve them in a herbarium. Botanical research took place mainly at the universities, and continued its traditional linkage with medicine. After the invention of the letterpress by Gutenberg in 1455, scientific literature proliferated and became accessible to a bigger audience. Plant species were reviewed and published rather quickly.

Herbalism, also known as botanical medicine, flourished in this period. The first herbal treatise to be published in English was the anonymous *Grete Herball* of 1526, and the two best-known herbals in English were *The Herball* or *General History of Plants* (1597) by John Gerard (1545–1612) and *The English Physician Enlarged* (1653) by Nicholas Culpeper (1616–1654), who was an English botanist, herbalist, and physician. The Age of Exploration introduced many new medicinal plants into Europe, and herbalism thrived on this.

Otto Brunfels (1488–1534), considered one of the three fathers of German botany along with Hieronymus Bock (1498–1554) and Leonhard Fuchs (1501–1566), was born in Braufels, near Mainz, Germany. He received a medical degree in 1531, and at a time when the study of medicine included the study of medicinal plants, Brunfels's interests were, not surprisingly, both medicine and botany. In 1530 he published his three-

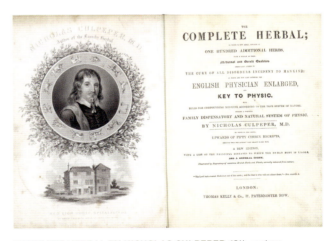

**COMPLETE HERBAL** BY NICHOLAS CULPEPER, 19th century

First published in 1653, the *Complete Herbal* contains descriptions of hundreds of medicinal herbs. Containing details such as where each herb can be found, its flowering time, astrology, and medical benefits, this historic and definitive guide to herbal remedies is still popular among herbalists.

**LEONHARD FUCHS**

Fuchs wrote *The History of Plants* in 1542, in which numerous plant species were described in detail. He described a genus of shrubs, which was named fuchsia in his honor. He also prepared the first important glossary of botanical terms, which helped pave the way for modern botany.

volume *Herbarum vivae eicones* (*Living images of Plants*). He later published another volume *Catalogus*, which included one of the first medical bibliographies. The *Herbarum* follows the tradition of earlier herbals, or illustrated books on plants that provided information on the medicinal uses of plants. It was only later that physicians and scholars developed interest in plants for their own sake.

Bock made extensive botanical excursions and collected plants that he cultivated in his private garden in order to study them. His work, called *A New Book on Herbs that Grow in German Countries with Special Regard Paid to Their Differences,* *Their Effects and Their Names,* was published in 1539. The second edition, published in 1546, was illustrated with 465 wood engravings and described plant structures and their vegetative development.

Fuchs, a bit of a chauvinist, wanted to re-establish the reputation of Greek scholars and end the intellectual dominance of the Arabs, especially in medicine and pharmacology. He recommended that in order to gain knowledge about plants, a study of nature was necessary. His work *De historia stirpium commentarii insignes* (*Notable commentaries on the history of plants*), published in 1542, belongs to the classical works of botanical literature and lists the plant species in alphabetical order. Fuchs deplored the widespread ignorance attached to medical practice in his time. Most of the practicing physicians used medicinal plants but had no knowledge of them. They relied for information on illiterate apothecaries, who in turn depended on peasants for gathering roots and herbs that were used in their study. Fuchs realized that patients could easily be poisoned rather than cured because of improper identification of plants. Therefore, he compiled a herbal dictionary to improve the German pharmacopoeia. While drawing upon the works of the Greek masters, Fuchs added at least 100 plants not mentioned in earlier herbals. He included many that had been introduced in 16th-century Germany from elsewhere, and he tried to ensure that all plants in his work were portrayed accurately.

During the 17th century, special attention was paid to German floras. Their names are often found in the families, generas, and species of modern nomenclature. G. Gesner (1516–1565) was the first to observe flowers and fruits closely and realize their value for classification. He made an attempt to divide flora into zones according to the altitudes where they were found, observing that the plants of the mountains differed from those of the plains in being small and sturdy.

Andrea Cesalpino (1519–1603), an Italian botanist, seems one of the first to study plants for their own sake, rather than for medical, decorative, or magical reasons. His *De plantis libri XVI* was the first work to consider issues like taxonomy, development, terminology, and nutrition in their own right, not as incidental while describing a specific plant. He developed a sophisticated taxonomy by using multiple characteristics. In delineating his taxa, he organized plants into families that correspond to the botanical families as we know them today. Groups like composites, grasses, and umbellifers are described with a high degree of precision. *De plantis* also contains a description of theoretical botany along with morphology, anatomy, biology, physiology, systematics, and nomenclature. He wrote that certain plants like hemp and nettle had male and female plants, and indicated that classification required not just the characteristics of the fruit had to be taken into account but also those of the flower and calyx. Cesalpino classified more than 1,500 plants and influenced later botanists profoundly. Theories of systematizing, naming, and classification dominated Europe throughout much of the 17th and 18th centuries.

Carolus Linnaeus (1707–1778) was a Swedish scientist who is best known for laying the foundations for the modern scheme of taxonomy. The Linnaean system classified living things within a hierarchy, starting with two kingdoms—the plant and animal kingdoms. Kingdoms were divided into classes, which were divided into orders, families, genera, and species. After Linnaeus's time, a few other ranks have been added, most notably phyla or divisions between kingdoms and classes. Groups of organisms are now called taxa or taxonomic groups.

Linnaeus's plant taxonomy was based solely on the number and arrangement of the reproductive organs. A plant's class was determined by its stamens (male organs) and its order by its pistils (female organs). This resulted in many groupings that seemed unnatural. For instance, Linnaeus's Class Monoecia, Order Monadelphia included plants with separate male and female "flowers" on the same plant (Monoecia) and with multiple male organs joined onto one common base (Monadelphia). This order included conifers such as pines, firs, and cypresses (the distinction between true flowers and conifer cones was not clear) as well as a few true flowering plants, such as the castor bean.

**CARL LINNAEUS,** Early 20th century French artwork

Linnaeus is considered the founder of modern taxonomy for he created a uniform system for classifying living organisms. In 1735, he published *Systema Naturae,* in which he classified flowering plants according to their sex organs. He introduced binomial nomenclature in his *Species Plantarum* (published in 1753), which is the denomination of each plant by two words, the genus name followed by the species name.

Plants without obvious sex organs were classified in the Class Cryptogamia, or "plants with a hidden marriage," which lumped together the algae, lichens, fungi, mosses, and other bryophytes, and ferns. Linnaeus admitted that this produced an "artificial classification," not a natural one, which would take into account all the similarities and differences between organisms. But like many naturalists of the time, Linnaeus attached great significance to sexual reproduction in plants, which had only recently been rediscovered.

Before Linnaeus, the nomenclature of species was disorganized. Many biologists gave their species long, unwieldy Latin names. A scientist comparing two descriptions of a species might not be able to tell which organisms were being referred to. For instance, the common briar rose was referred to by different botanists as *Rosa sylvestris inodora seu canina* and as *Rosa sylvestris alba cum rubore, folio glabro*. The import of new plants and animals that were being brought back to Europe from expeditions in Asia, Africa, and the Americas forced the need for a workable, standard system of naming species. Linnaeus simplified this by establishing binomial nomenclature, that is, naming plants by designating one Latin name to indicate the genus and one name for the species. The two names make up the binomial species name. For instance, in his two-volume work *Species Plantarum* (*The Species of Plants*), Linnaeus renamed the briar rose *Rosa canina*.

This binomial system became the standard system for naming species. The oldest plant names accepted as valid today are those published in *Species Plantarum* in 1753, while the oldest animal names are those in the tenth edition of *Systema Naturae* (1758), the first edition to use the binomial system consistently throughout. Among Linnaeus's other important works are: *Fundamenta botanica* (1737), *Bibliotheca botanica* (1736), *Flora lapponica* (1737), *Hortus Cliffortianus* (1737), *Critica botanica* (1737), *Flora svesica* (1745), and *Philosophica botanica* (1751).

The British researcher John Ray (1627–1705) established six rules of plant classification in 1703, which belong to the fundamental principles of plant systematics even today:

- To avoid confusion and errors, names should not be changed.
- Characteristics have to be exactly and distinctively defined. Those based on relative properties like height are not to be used.
- Everybody should easily detect characteristics.
- Groups accepted by almost all botanists should be kept.
- Care has to be taken that related plants are not separated, and "unnatural" ones and those that are different are not to be united.
- If not necessary, characteristics should not be increased in number. Only as many should be used as are necessary to make a reliable classification.

**JOHN RAY,** The National Portrait Gallery, London

Ray collected the specimens shown below from Devil's Dike, near Cambridge, in the early part of his career as a naturalist. He divided the plants into three taxa: the cryptogams or flowerless plants, the monocotyledons, and the dicotyledons—the basis of the system still used today. Ray's work covers the taxonomy of about 18,600 species.

The English botanist Nehemiah Grew (1641–1712) made significant contributions to plant anatomy. In 1672 he published *The Anatomy of Vegetables Begun*, followed in 1682 by *The Anatomy of Plants*. Grew recognized cells in plants, but their biological significance evaded him. He recognized flowers as the sexual organs of plants and described their parts. He also described the individual pollen grains and observed that they are transported by bees. Grew postulated that the pollen contained in the stamen was necessary for pollination and that tissues made up of cells were the basic elements of vegetative matter. He examined the pith of the stem, and distinguished three different types of fibers in it: single fibers, screw-like fibers, and sap-containing fibers of the bast. Grew described the development of the wood and the arrangement and shape of stomata as well as the parenchyma. Twelve years after the publication of *The Anatomy of Plants*, a German physician utilized Grew's anatomical studies in experiments to verify sexual reproduction in plants.

## ZOOLOGY AND ANATOMY

In the early 16th century, scholars from various universities, such as Bologna and Paris, returned to the works of Galen and emphasized his superiority over his later interpreters, stressing the importance of anatomy in medicine. Galen had combined the philosophical work of Aristotle and other Greeks with his own lifetime of dissections, creating a system that explained not just the structure of the human body but also how the body worked.

After the fall of Rome, Galen's legacy lived on in Arab cities like Baghdad, where his work was translated and interpreted. In the 1100s, Europeans began to translate Galen from Arabic and made his work the basis of medical training. But in the many steps of translation, much of the spirit of Galen's work—especially his emphasis on observing for oneself rather than relying on authority—was lost.

**ANDREAS VESALIUS**

Vesalius was a professor of anatomy and surgery at Padua university, Italy. In 1543 he published *De Humani Corporis Fabrica*, which set a completely new level of accuracy in anatomy. His book was controversial because it disagreed with the views of Galen, the accepted medical authority of the time. *Fabrica* was rich in illustrations of the whole body and details of it, and acquired a reputation for its great beauty.

Andreas Vesalius (1514–1564) was a Belgian anatomist and physician whose dissections of the human body and descriptions of his findings helped to correct many misconceptions that had been carried forward from ancient times. The most significant was his correction of Galen's theories. During his research, Vesalius showed that the anatomical teaching of Galen, revered in medical schools, was based upon the dissections of animals but were meant as a guide to the human body. Vesalius wrote the highly accurate and comprehensive anatomical texts called *De Humani Corporis Fabrica* (*On the Structure of the Human Body*), which were seven volumes on the structure of the human body. *Fabrica* launched a new tradition in anatomy in Europe, in which anatomists began to trust only their own observations. Vesalius's discovery of the important differences between species paved the way for the science of comparative anatomy, in which researchers studied animals to find their similarities and differences. In the process, they gradually began to recognize humans as being one species among many, with a few unique traits but many others shared in common with other animals.

**WILLIAM HARVEY,** Oil on canvas, by Robert Hannah, 1848

Harvey was an English physician who was the first to describe accurately how blood was pumped around the body by the heart. He is explaining his discovery of the circulation of the blood to King Charles I in the above 19th century painting.

The English physician William Harvey's (1578–1657) understanding of the circulation of blood places him among the most accomplished physicians and scientists of his time. Harvey also pioneered the study of embryology and experimental physiology. The Roman physician Galen had postulated that the liver received food from the small intestine and converted it to blood. The heart pumped this blood to the other organs, and those organs consumed it. This belief lasted more than 1400 years till Harvey disproved it experimentally. He measured the amount of blood pumped by the hearts of snakes and other animals, and concluded that the heart pumps more blood in half an hour than there is in the entire body; that animals do not consume enough food to account for the blood in their body; and that the blood must be continually circulated around the body. This last conclusion was based on his belief that the human body is modeled after the solar system, and, like the planets orbit the sun, the blood goes round and round, pumped by the heart. Harvey predicted that there must be a connection between the arteries and veins to allow the blood to get back to the heart. This connection turned out to be the capillaries, which were first seen by Antony van Leeuwenhoek and Marcello Malpighi after Harvey's death.

Harvey laid the foundation of embryology by saying that every animal, including humans, arises from the union of sperm and egg, thus challenging the prevalent notion of spontaneous generation according to which animals arise from decaying flesh. In 1651 he published this theory of animal development, with a detailed account of the embryology of the chick, as *Studies on the Generation of Animals*. Underscoring the basic principle of embryology, his book bore the inscription *ex ovo omnia* (or, everything comes from an egg).

## MICROBIOLOGY

Microbiology as a branch of biology developed in the 17th and 18th centuries after crude microscopes were created in the late 16th century. After its invention, the microscope became a much used tool of research. P. Borelli from the Netherlands detected nerves, spots, and star-shaped hairs when he observed leaves. Robert Hooke (1635–1703) and Nehemiah Grew of England, Marcello Malpighi (1628–1694) of Italy, and the Dutchmen Anton van Leeuwenhoek (1632–1723) and Jan Swammerdam (1637–1680), all outstanding microscopists of the 17th century, could be considered the founding fathers of microbiology.

Leeuwenhoek of Holland is credited with making some of the most important discoveries in the history of biology because of his success in making and using microscopes that Zacharias Janssen (1580–1638) had invented in 1590. Leeuwenhoek is known to have made over 500 microscopes, which were powerful magnifying glasses, arranged in an instrument about

3–4 inches (7–10 cm) long, which had to be held up close to the eye. Leeuwenhoek's curiosity led him to study as many things as possible under his home-made microscopes. With his shaving razor he would slice off very thin slices of cork, plants, or other specimens to view under his microscopes. Using his handcrafted microscope, Leeuwenhoek was the first to observe and describe muscle fibers, bacteria, spermatozoa, and blood flow in capillaries.

Leeuwenhoek collected water samples from a lake to observe under his microscope and made excellent descriptions of the life forms he saw: "Passing just lately over this lake, . . . and examining this water next day, I found floating therein divers earthy particles, and some green streaks, spirally wound serpent-wise, and orderly arranged, after the manner of the copper or tin worms, which distillers use to cool their liquors as they distil over. The whole circumference of each of these streaks was about the thickness of a hair of one's head . . . all consisted of very small green globules joined together: and there were very many small green globules as well." In 1702, Leeuwenhoek described protists, including the ciliate Vorticella: "In structure these little animals were fashioned like a bell, and at the round opening they made such a stir, that the particles in the water thereabout were set in motion thereby . . ." Leeuwenhoek looked at animal and plant tissues, mineral crystals, and fossils. He discovered bacteria, free-living and parasitic microscopic protists, sperm cells, blood cells, microscopic nematodes and rotifers, and much more. His discoveries, which were widely circulated, opened up an entire world of microscopic life to scientists.

As a member of the Royal Society, he wrote approximately 560 letters to the society and other scientific institutions over a period of about 50 years. These letters dealt with the subjects he had investigated. In 1674 he discovered infusoria, two years later he discovered bacteria, the following year spermatozoa, and in 1682 he discovered the banded pattern of muscular fibers.

### ANTON VAN LEEUWENHOEK

**Far Left:** Leeuwenhoek discovers the microbe.

**Left:** The first microscope made by Leeuwenhoek. The Dutch scientist's microscopes used a single lens, which he ground to near perfection himself. The lens was clamped between two brass plates with holes for the viewer to look through, and the specimen was mounted on the tip of an adjustable pointer. Magnifications of around x200 could be achieved—enough for Leeuwenhoek to become the first to see protozoa (which he called "little animalcules"), human sperm (a discovery he reported somewhat nervously), and even, in 1683, bacteria.

The Italian physiologist Marcello Malpighi (1628–1694), while observing dissected lung tissue, discovered a network of tiny thin-walled microtubules on the surface of the lung and of the distended bladder of the frog, which he named capillaries. He said that capillaries were the connection between arteries and veins that allowed blood to flow back to the heart. Although Harvey had correctly inferred the existence of capillary circulation, he had never seen it.

In 1669, Malpighi published his work on the silkworm showing that silkworms have no lungs, and breathed through a row of holes located on the side of their bodies which connected a system of tubules that he termed trachea. He thought that plants and animals had similar breathing mechanisms, erroneously believing that tiny tubes found in many plants performed the same function as did trachea in insects.

Malpighi used the microscope for studies on skin, kidney, and for the first interspecies comparison of the liver. His name is associated with the soft or mucous character of the lower stratum of the epidermis, the vascular coils in the cortex of the kidney, and the follicular bodies in the spleen. Malpighi also extended the science of embryology. The use of microscopes enabled him to describe the development of the chick in its egg. Malpighi discovered the taste buds while examining human tongues, and recognized that the liver functioned as a gland. He was the first to discover and study human fingerprints, and the first to attempt the finer anatomy of the brain. His descriptions of the distribution of gray matter and of the fiber tracts in the cord, with their extensions to the cerebrum and cerebellum, were accurate, but he went wrong in interpreting the gray matter as a glandular structure that secreted the "vital spirits."

**MARCELLO MALPIGHI,** Engraving from an oil painting by A.M. Tobar

Malpighi was the first to build a microscope, slightly ahead of Leeuwenhoek. He was a physician by training, graduating at Bologna in 1653. Malpighi's observations through his microscope were to revolutionize anatomy. He showed a mechanism for the transfer of air into the blood in the lungs, and observed capillary vessels for the first time. This completed the theory of blood circulation advanced by Harvey.

**Below:** Drawing of a chick's embryo in Malpighi's *On the Formation of the Chick in the Egg*, taken from the 1686 edition.

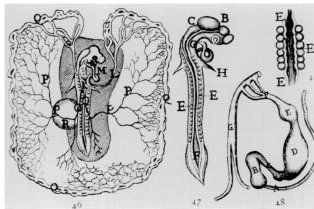

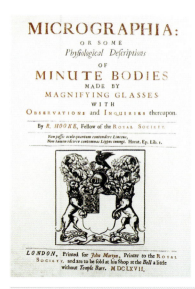

**HOOKE'S ILLUSTRATION OF CORK WOOD CELLS**

This drawing of cork, as seen under a microscope, is the first picture showing "cells" in a biological specimen. The illustration appeared in his book *Micrographia* (left), which was published in 1667 and is believed to be the first major book on microscopy.

**HOOKE'S MICROSCOPE,** *Micrographia*, 1665

Hooke devised the compound microscope and illumination system shown above, one of the best such microscopes of his time. With it, he observed organisms as diverse as insects, sponges, bryozoans, foraminifera, and bird feathers.

Jan Swammerdam was a Dutch naturalist trained as a physician at the University of Leiden. In the 1660s, Swammerdam made major discoveries in anatomy where he demonstrated the presence of valves in lymph vessels. His work on the frog muscle demolished the pre-scientific ideas that nervous action was due to "vital spirits." He discovered the mechanism of penile erection and was one of the first to discover the human ovarian follicles.

In 1669, Swammerdam wrote the *Historia Insectorum Generalis* (*The Natural History of insects*), in which he proposed a classification of insects based on their modes of development. This is still followed today. He identified four "orders" of insects, corresponding to different developmental pathways. The first, and simplest, is a diverse groups of creatures such as spiders, scorpions, snails, and ametabolous insects (such as lice) where the adult form hatches directly out of the egg. In the second order, which includes dragonflies, locusts, and the mayfly, a nymph hatches out of the egg and then gradually develops into the adult form. The third and fourth orders—holometabolous

insects with a pupal stage, such as butterflies, bees, or flies—are those that posed the greatest intellectual challenge at the time and remain poorly understood even today. In addition to his work on insect taxonomy, Swammerdam's study showed, once and for all, that there was no such thing as spontaneous generation. He demonstrated that insects did not generate spontaneously but were the product of an egg laid by a female of the same species. He also showed that the various stages in the insect's life-cycle—egg, larva, pupa, and adult—merely represent different forms of the same individual. Swammerdam noticed differences between males and females of the same species. In 1670, inspired by Malpighi's study of the silkworm, Swammerdam began to dissect insects and noted observations on insect anatomy and development, including a study of the bee. These studies have stood the test of time and still provide valuable information and insights.

Robert Hooke, an Englishman, was a highly gifted experimental scientist of the 17th century, with diverse interests, ranging from physics and astronomy, to chemistry, biology, and geology.

Hooke's major contribution to biology is his book *Micrographia*, published in 1665. Here he recorded his observations with the compound microscope of insects, sponges, bryozoans, foraminifera, and bird feathers. *Micrographia* was an accurate and detailed record of his observations, illustrated with magnificent drawings. Hooke's most famous microscopical observation was his study of thin slices of cork. In "Observation XVIII" of the *Micrographia*, he wrote: ". . . I could exceedingly plainly perceive it to be all perforated and porous, much like a Honey-comb, but that the pores of it were not regular . . . these pores, or cells . . . were indeed the first microscopical pores I ever saw, and perhaps, that were ever seen, for I had not met with any Writer or Person, that had made any mention of them before this." What Hooke had in fact seen were the cell walls in cork tissue, and he went on to coin the term "cells" because the boxlike cells of cork reminded him of the cells of a monastery. Hooke also reportedly saw similar structures in wood and in other plants. In 1678, after Leeuwenhoek had written to the Royal Society about his discovery of bacteria and protozoa, Hooke was asked by the Society to confirm Leeuwenhoek's findings. He did so, thus helping to establish the field of microbiology.

**DRAWING OF NETTLE,** From *Micrographia*

This illustration of the underside of a stinging nettle leaf was printed in Hooke's *Micrographia*. Hooke correctly discovered that the stinging hairs were hollow "from top to bottom." He did this by observing the movement of the liquid within the sting into his finger while using the microscope. Somewhat fancifully, he compared the base of the sting to a leather bag, and the sting itself to a glass tube. Besides this, *Micrographia* contains some of the most beautiful drawings of microscope observations ever made.

Hooke was the first person to examine fossils with a microscope and began the study of paleontology. He noted similarities between the structures of petrified wood and fossil shells on the one hand, and living wood and living mollusc shells on the other. In *Micrographia*, he compared a piece of petrified wood with a piece of rotten oak wood, and concluded that dead wood could be turned to stone by the action of water rich in dissolved minerals, which would deposit minerals throughout the wood. Hooke also mentions in *Micrographia* that the shell-like fossils he examined really were "the shells of certain shel-fishes, which, either by some deluge, inundation, earthquake, or some such other means, came to be thrown to that place, and there to be fill'd with some kind of mud or clay, or petrifying water, or some other substance . . ."

Hooke had correctly understood that fossils are not some kind of natural stones with extraordinary designs but remains of living organisms. Hooke realized, 250 years before Darwin, that fossils document changes in the organisms found on the planet, and that species have both appeared and gone extinct throughout the history of life on Earth.

It was only in the 19th century that cells were recognized as the basis of life. Besides his microscopical studies, Hooke worked on physiological processes and postulated that the snapping of the leaf of *Mimosa pudica* was caused by the excretion of a very delicate liquid. He explained that the stinging of nettles was due to the flow of a caustic sap out of the bristles of the plant. The microscope-based scientists understood that the picture seen in the microscope was only a part and that a whole image had to be gained out of all the partial ones by connecting

## JEAN LAMARCK

Lamarck's theory of evolution, known as Lamarckism, was based on the idea that acquired characteristics are inherited. After 1800 he put forward general ideas on plant and animal species, which he believed were not "fixed." He proposed that in nature it is the environment that produces change; the length of the giraffe's neck, for example, he attributed to generations of reaching up for food. Lamarckism was largely abandoned after the work of Darwin and Mendel.

them. They also understood that the examinations had to be carried out with the aim to grasp the whole inner structure of the plant.

Linnaeus also contributed to the field of zoology. He proposed that there were four subcategories under Homo sapiens: Americanus, Asiaticus, Africanus, and Europeanus. These categories were based on place of origin at first, and later skin color. Each race had certain characteristics that were endemic to members of that race. According to him, Native Americans were reddish, stubborn, and angered easily. Africans were black, relaxed, and negligent. Asians were sallow, avaricious, and easily distracted. Europeans were white, gentle, and inventive. Over time, this classification led to a racial hierarchy, in which Europeans were at the top.

Jean-Baptiste-Pierre-Antoine de Monet Lamarck (1744–1829) believed that organisms are not passively altered by their environment but that a change in the environment causes changes in the needs of organisms living in that environment, leading to a change in their behavior. Altered behavior causes greater or lesser use of a given structure or organ, overuse would cause the organ to increase in size over several generations, whereas disuse would cause it to shrink or even disappear. Lamarck called this rule—that use or disuse causes structures to enlarge or shrink—the first law in his book *Philosophie zoologique* (*Zoological Philosophy*). Lamarck's second law stated that all such changes were heritable. The result of these laws was to establish the continuous, gradual change of all organisms, as they became adapted to their environments.

Lamarck's *Philosophie zoologique* mentions the great variety of animal and plant forms produced under human cultivation; the presence of vestigial, non-functional structures in many animals; and the presence of embryonic structures that have no counterpart in the adult. Like Darwin and later evolutionary biologists, Lamarck argued that the earth was immensely old. He even mentions the possibility of natural selection in his writings, although he never seems to have attached much importance to this idea. The overarching component of Lamarckian evolutionism was what became known as the inheritance of acquired characters. This described the means by which the structure of an organism altered over generations. Depending on its survival needs, an animal may undergo physiological changes in its lifetime, which it passes on to its offspring.

According to Lamarck, evolution was a complex process not driven by chance. He wrote in *Philosophie zoologique*, "Nature, in producing in succession every species of animal, and beginning with the least perfect or simplest to end her work with the most perfect, has gradually complicated their structure." Lamarck did not believe in extinction: for him, species that disappeared did so because they evolved into different species.

# Chapter 3
# MODERN BIOLOGY

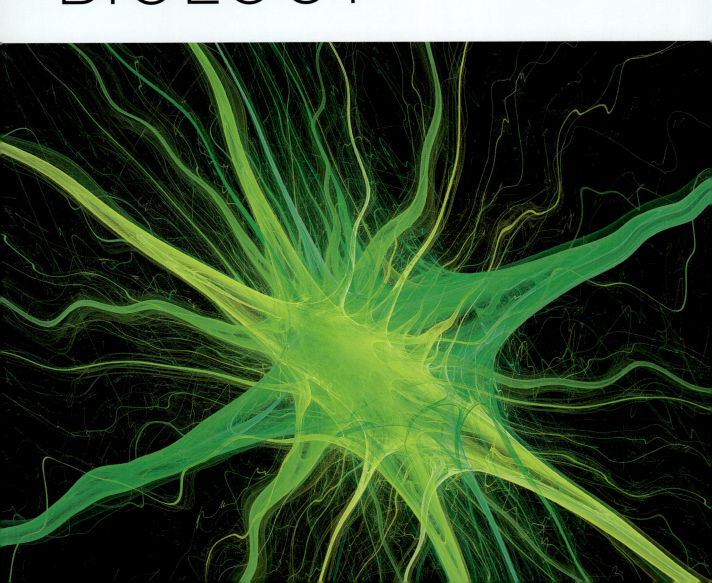

# PHYSIOLOGY

Physiology is the study of the physical and chemical processes that take place in living organisms during the performance of life functions. It is concerned with such basic activities as reproduction, growth, and metabolism as they are carried out within the cells, tissues, and organs of the body. Physiology is linked to anatomy and was historically considered a part of medicine. There are three broad divisions in it: general physiology, concerned with basic processes common to all life forms; the physiology and functional anatomy of humans and animals; and plant physiology. The first studies in human physiology were probably undertaken about 300 BC by the Alexandrian physician Herophilus, who dissected the bodies of criminals.

Modern animal physiology dates from the discovery of the circulation of blood by William Harvey in 1616. Thereafter, the Flemish chemist Jan Baptista van Helmont (1580–1644) developed the concept of gases and suggested the use of alkalies in treating digestive disturbances. The Italian Giovanni Alfonso Borelli (1608–1679) described animal motion, suggesting that the basis of muscle contraction lay in the muscle fibers. After the work of Leeuwenhoek and Malpighi, Thomas Wharton (1614–1673) demonstrated salivary secretion and Nicolaus Steno (1638–1686) demonstrated the secretions of the tear glands and salivary glands. The Dutch physician Regnier de Graaf (1641–1673) discovered the follicles in the ovary and studied pancreatic juices and bile. The English physician Richard Lower (1631–1691) was the first to transfuse blood from one animal to another, and the French physician Jean Baptiste Denis (1643–1704) first gave a human being a successful blood transfusion.

In the 17th century, advances were made in the study of respiration. The English physiologist John Mayow (1645–1679) showed that air was not a single substance but a mixture of several, not all of which were necessary for life. In the 18th century, the British chemist Joseph Priestley (1733–1804) showed that the proportion of oxygen essential for animal life is identical with the proportion of oxygen needed to support combustion. Antoine-Laurent Lavoisier (1743–1794), the French chemist, isolated and named oxygen shortly thereafter, and showed that the by-product of respiration is carbon dioxide.

There were important developments in plant physiology during the 16th century. In plants, the circulation of sap, the transformation of soil compounds into plant-specific components, the connection of water uptake, sap pressure and evaporation, and single plant substances began to be studied. C. Perrault (1613–1688) made extensive observations of the circulation of sap in plants. He believed the "fermentation of the soil humidity" and the effects of the sap in the roots to be the reason for the rising of the sap in a plant. This created warmth in the plant, expanding the sap. He explained the descending movement of the sap by the growth of the roots and an interaction between leaves and roots.

In 1679, E. Mariotte (1620–1684) recognized that the most dissimilar plants can receive their nutrition from the same components of the soil and that they are able to form many more compounds than are to be found in the soil. In addition, he observed that the same sap can produce sour fruit in a wild pear tree and sweet juicy ones in good stock grafted to the wild tree. By distilling the sap of different plants he showed that the same species always contains the same compounds.

John Woodward (1665–1728) of Britain carried out a series of experiments on plant nutrition from 1696. In 1699, he published his essays on transpiration. He showed that the greater part of the water absorbed by a growing plant is exhaled through its pores into the atmosphere, that in three months the plant emits 46 times the amount of water that it stores in itself. C. W. Scheele (1742–1786) and Louis Nicolas Vauquelin (1763–1829) studied single plant substances like tartaric acid, citric acid, malic acid, oxalic acid, and gallic acid. A. S. Markgraf (1709–1782) from Berlin and Duhamel du Monceau (1700–1782) analyzed plant ashes and identified a number of salts in them.

During the second half of the 18th century, the Italian physician Luigi Galvani (1737–1798) showed that the muscles of a frog's leg could be made to contract by stimulation with an electric current. The Italian physiologist Lazzaro Spallanzani (1729–1799) investigated the activity of gastric juice in digestion and studied fertilization and artificial insemination in lower animals. In the 19th century, Claude Bernard (1813–1878) investigated carbohydrate metabolism in humans and described the functions of the autonomic nervous system. According to Bernard, the basis of health is an organism's success in maintaining internal equilibrium.

**LUIGI GALVANI'S EXPERIMENT WITH FROG MUSCLES,** From a children's book on electricity, c. 1900

During the 1780s, Galvani noticed that the muscles of dissected frog legs twitched wildly when struck by an electric spark. He called this animal electricity, the life force within the muscles of the frog.

In the 20th century, Walter Bradford Cannon (1871–1945) named the dynamic equilibrium state "homeostasis" and showed that the body could adjust to meet serious external danger. Cannon demonstrated such processes of the human body as internal regulation of body heat, alkalinity of the blood, and preparation of the body for defense by the secretion of epinephrine, also called adrenaline, in the adrenal gland. Among the most important advances of the 20th century are the discovery of new hormones, the recognition of the role of vitamins, the discovery of blood types, and the development of the electrocardiograph to record the activity of the heart and electroencephalograph to record the activity of the brain. In this period, advances have been made in understanding metabolism, the role of enzymes, and the immune system.

The 20th century also witnessed fundamental advances in neurology. Edgar Douglas Adrian (1889–1997) measured and recorded electric potentials from sense organs and motor nerve fibers and Charles Scott Sherrington (1857–1952) investigated the integrative action of the nervous system. Following this, Joseph Erlanger (1874–1965) and Herbert Spencer Gasser (1888–1963) recorded the variation of electrical impulses that occurs in the nerve fibers. Later investigations by the American biochemist Julius Axelrod (1912–2004), the Swedish physiologist Ulf von Euler (1905–1983), and the British physician Bernard Katz (1911–2003) demonstrated the role of neurotransmitters, the specific chemicals that mediate transmission of nerve impulses.

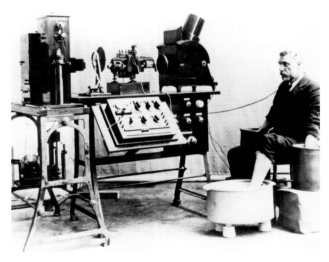

**AN EARLY ELECTROCARDIOGRAPH (ECG)**

A patient is seen using an early ECG to take a measurement from his skin of the electrical signals in his heart. The patient's left arm and leg are placed in buckets of salt water to improve the electrical contact.

**1800** The term biology in its modern sense is coined by Karl Friedrich Burdach.

**1817** Pierre-Joseph Pelletier and Joseph-Bienaime Caventou isolate chlorophyll.

**1838** Matthias Schleiden proposes that all plants are composed of cells.

**1859** Darwin publishes *The Origin of Species*.

**1866** Gregor Mendel formulates his Laws of Inheritance.

**1869** Friedrich Miescher discovers nucleic acids in the nuclei of cells.

**1886** Scientists explain the nitrogen-fixing of the pea family.

**1898** Martinus Beijerinck uses filtering experiments to show that tobacco mosaic disease is caused by something smaller than a bacterium, which he names a virus.

**1918** Hermann Muller formulates the chief principles of spontaneous gene mutation.

**1920s** Nucleic acid found to be a major component of the chromosomes.

**1928** Frederic Clements proposes the theory of plant succession.

**1944** Oswald Avery identifies nucleic acids as the active principle in bacterial transformation.

**1946** American chemist Melvin Calvin explains photosynthesis.

**1952** Alfred Hershey and Martha Chase show that DNA, and not proteins, are responsible for the transmission of gene information.

**1953** Watson and Crick determine that deoxyribonucleic acid (DNA) is a double-strand helix of nucleotides.

**1969** Robert Whittaker proposes five kingdoms of life.

**1986** E.O. Wilson coins the term biodiversity.

**2000** The Human Genome Project presents its preliminary results: each of the body's 100 trillion cells contains some 3.1 billion nucleotide units. Only 1 per cent of these are thought to be transcriptional, clustered in possibly as few as 30,000 genes.

# BIOCHEMISTRY

Biochemistry is the study of the chemical processes that take place within living organisms. As its name suggests, it combines the study of both chemistry and biology. For a long time, it was thought that only living organisms could create special biological molecules from other biological molecules obtained through food. These molecules were thought to be imbued with a "vital force" that made life possible. In 1828, the German chemist Friedrich Wöhler (1800–1882) disproved the notion of vital force by accidentally synthesizing the organic chemical urea—a major component of urine—from inorganic precursors.

**FRIEDRICH WÖHLER**

Studies in biochemistry have elucidated diverse phenomena like the process of photosynthesis, the conversion of glucose to energy, the production of lactic acid when muscles are exercised, and how proteins are synthesized in the cell. The molecules studied by biochemists include carbohydrates, proteins, lipids, and nucleic acids. Most of these have a structural function, like fats, carbohydrates, and proteins. The fat in our body is made of lipids which are non-water-soluble biomolecules. Based on genetic instructions proteins are manufactured directly and are among the most complex organic molecules. Nucleic acids are the building blocks of our genetic instructions (DNA and RNA) found in all forms of life, from humans to viruses. The distinct pattern of nucleic acids found in the nuclei of a species' cells is called its "genome."

Another branch of biochemistry developed from agricultural research, and studied nutrition, especially the role of vitamins in health. Diseases identified as resulting from vitamin deficiencies in the diet, such as rickets, scurvy, and pellagra, were cured and prevented by specific dietary changes.

During World War II, American biochemists were involved in the large-scale production of penicillin, other antibacterial drugs, and blood fractionation products for use in transfusion. These projects involved complex interactions between basic and applied research, managed through close collaborations between scientists, government, and industry, which led to massive expansion of public funding for basic biochemical research in postwar America.

Biochemists have been closely involved with the human genome project. The current research on stem cells has led to very important information about chemical processes that cause cell death. If stem cells are to be used to repair parts of the body, it is essential that they remain viable. Understanding chemical signals that might kill a population of stem cells aids in the understanding of when and how stem cells might be used.

Biotechnology, the use of living things to make products, is another field in which the biochemistry expert thrives. In studies regarding food, biochemists might work in a number of practical ways, such as product development of foods that are least likely to cause weight gain, or developing foods that have beneficial qualities. In most wineries and breweries, biochemistry is applied to evaluate yeasts and acids used to make alcohol.

The pharmaceutical industry greatly relies on biochemistry because the chemical make-up of the body must be studied in relationship to the chemicals we might put in our body as medicines. Certain medications have been developed directly as a result of biochemistry research. Antidepressants like Prozac, called serotonin reuptake inhibitors, are used because there is an underlying medical assumption that in depressed people serotonin gets used too quickly by the body, affecting the mood significantly. By inhibiting the body's utilization of serotonin, more free serotonin is allowed to circulate and thus improve the mood.

Biochemistry helps to make the development of drugs like Prozac possible because theories based on these drugs derive specifically from the study of chemicals produced by the body that affect mood. Biochemistry work in hormones, enzymes, proteins, and cell interaction, all enhance the understanding of what type of chemicals might be needed to correct imbalances, without adversely affecting the other chemicals produced in the body.

## CELL THEORY

The invention of the microscope in the 17th century allowed the study of life forms invisible to the naked eye. The Jesuit priest Athanasius Kircher (1601–1680) showed in 1658 that maggots and other living creatures developed in decaying tissues. In the same period, oval red-blood corpuscles were described by Swammerdam, who also discovered that a frog embryo consists of globular particles. But the first description of the cell is attributed to Robert Hooke. In his *Micrographia*, the first

important work devoted to microscopical observation, Hooke described the microscopic units that made up the structure of a slice of cork and coined the term "cells" or "pores" to refer to these units.

The cell theory, which states that the cell is the basic component of living organisms, was formulated in 1838. Earlier cells were not seen as undifferentiated structures, but some components, such as the nucleus, had already been visualized before 1838. The Scottish botanist Robert Brown (1773–1858) was the first to recognize the nucleus, a term that he introduced, as an essential constituent of living cells. The botanist Matthias Jakob Schleiden (1804–1881) suggested in 1838 that every structure of plants is composed of cells or their products. The following year, a similar conclusion was elaborated for animals by the zoologist Theodor Schwann (1810–1882). He stated that "the elementary parts of all tissues are formed of cells" and that "there is one universal principle of development for the elementary parts of organisms . . . and this principle is in the formation of cells." The conclusions of Schleiden and Schwann are considered to represent the official formulation of the "cell theory." In the 1850s, Robert Remak (1815–1865), Rudolf Virchow (1821–1902) and Albert Kölliker (1817–1905) showed that cells are formed through the division of pre-existing cells. Virchow's aphorism *omnis cellula e cellula* (every cell from a pre-existing cell) became the basis of the tissue formation.

After Schleiden and Swann's formulation of the cell theory, the basic constituents of the cell were considered to be a wall or a simple membrane, a viscous substance called "protoplasm" (later replaced by the term "cytoplasm"), and the nucleus. By

**THEODOR SCHWANN (FAR LEFT) AND MATTHIAS SCHLEIDEN**

Schwann and Schleiden co-founded the cell theory of life, that all animal tissues were made from tiny cells. Plant cells had been observed two centuries earlier by Robert Hooke and others.

# THE NEURON THEORY

Because of its softness and fragility, nervous tissue was difficult to handle. It was also structurally complex and could not be fitted into the cell theory as understood then. Nerve-cell bodies, prolongations, and nerve fibers were seen in the first half of the 19th century but defied a three-dimensional reconstruction.

In 1872, the German histologist Joseph Gerlach (1820–1896) proposed that in the central nervous system, nerve cells established anastomoses with each other through a network formed by the minute branching of their dendrites. According to this concept, the network or reticulum was an essential element of gray matter that provided a system for anatomical and functional communications, a protoplasmic continuum from which nerve fibers originated. Another important breakthrough in neurocytology and neuroanatomy came in 1873 when Italian histologist Camillo Golgi (1843–1926) developed the black reaction which allowed him to observe the branching of the axon. Gerlach and Golgi gave the hypothesis that the nervous system represented an exception to the cell theory, being formed not by independent cells but rather by a gigantic inter-connected network.

Later, the Spanish neuroanatomist Santiago Ramón y Cajal (1852–1934)

**CAMILLO GOLGI (TOP)**

Golgi developed a staining agent based on silver compounds in 1873, which made it possible to observe the paths of nerve cells in the brain for the first time.

became the main supporter of the neuron theory. His neuroanatomical investigations contributed to the foundations of the basic concepts of modern neuroscience. However, definitive proof of the neuron theory was obtained only after the introduction of the electron microscope, which allowed identification of synapses between neurons. In this way, the nervous system, found to be made up of independent units, became integrated in the cell theory.

**SANTIAGO RAMÓN Y CAJAL CHARTING THE NERVOUS SYSTEM (BELOW)**

Cajal used a histological staining technique developed by his contemporary Golgi to study the structure of the central nervous system.

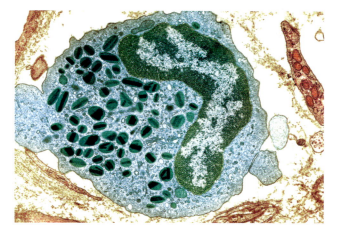

plague to microorganisms. However, most scientists in Europe believed that disease-causing germs arose from spontaneous generation, that flies, worms, and other such creatures appeared to arise spontaneously from decaying matter. Aristotle believed this, and the idea persisted into the 19th century.

At this time, the great French chemist Louis Pasteur (1822–1895) conducted seminal experiments that conclusively overturned the idea that life could be spontaneously generated. He demonstrated that there are microorganisms everywhere, including in the air and that they were the source of the chemical process by which milk soured. The process he developed that heats milk (and other liquids) to kill the microbes is called pasteurization. After it was widely adopted, pasteurization

the end of the 19th century, the principal organelles that are now considered to be parts of the cell were identified. The term "ergastoplasm" (endoplasmic reticulum) was introduced in 1897, and mitochondria were observed by several scientists and named by Carl Benda in 1898.

Like the protoplasm, the nucleus too had an uneven appearance and the nucleolus. A number of structures appeared during cell division in the shape of ribbons, bands, and threads. As these structures could be heavily stained, they were called "chromatin" by Walther Flemming (1843–1905), who also introduced the term "mitosis" in 1882 and gave a description of its processes. Flemming observed the longitudinal splitting of salamander chromosomes during the metaphase and established that each half chromosome moves to the opposite pole of the mitotic nucleus. This process was also observed in plants, providing further evidence of the unity of the living world.

## GERM THEORY

The earliest mention of "minute creatures" that cause disease is found in the ancient Indian Vedas. The 11th-century Persian polymath Avicenna understood that tuberculosis was contagious. Later Muslim scholars attributed the bubonic

**LOUIS PASTEUR INVESTIGATING FERMENTATION**

Pasteur studied the process of souring of wine and beer. He was able to demonstrate that fermentation does not require oxygen, but it involves living organisms. He recognized that there are two kinds of yeast, one of which produces alcohol and the other lactic acid, which turns wine sour. He suggested that heating wine or beer at about 120 degrees Fahrenheit would kill the yeast, preventing further fermentation and souring due to lactic acid production. This heating process used to kill undesirable microorganisms is known as pasteurization and is also applied to milk.

ensured that milk ceased to be a source of tuberculosis and other diseases. In his experiment, Pasteur filled short-necked flasks with beef broth and boiled them, sealing some and leaving others open. He was able to show that while the sealed flasks remained free of microorganisms, the open flasks were contaminated within a few days. In a second set of experiments, Pasteur placed broth in flasks that had open-ended, long necks. He boiled the broth after bending the necks of the flasks into S-shaped curves that dipped downward and then swept sharply upward. The contents of these uncapped flasks remained uncontaminated even months later. Pasteur explained that the S-shaped curve allowed air to pass into the flask; however, the curved neck trapped airborne microorganisms at the bottom of the curve, preventing them from traveling into the broth.

Pasteur believed that microorganisms were responsible for infectious diseases in humans and animals, and for their transmission among them. He developed effective vaccines against anthrax and rabies by harvesting tissue from animals that had died from these diseases. However, it was Robert Koch (1843–1910) who finally validated the germ theory of disease. He identified the specific bacteria that caused anthrax, tuberculosis, and cholera, and developed a set of rules, called Koch's Postulates, for determining conclusively whether a microorganism is the source of a disease. With this, the science of bacteriology was born.

## THE THEORY OF EVOLUTION

In 1859, Charles Darwin (1809–1882) revolutionized the zoological and botanical sciences by introducing the theory of evolution by natural selection as an explanation for the diversity found in animals and plants. Almost a thousand years before Darwin, the Arab scholar Al-Jahiz had already developed a rudimentary theory of natural selection. Al-Jahiz described the struggle for existence in his *Book of Animals*,

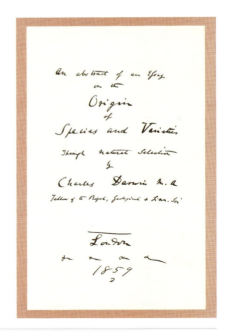

**THE ORIGIN OF SPECIES,** Title Page Prospectus in Darwin's own hand, 1859

In this book Darwin first expounded upon the theory of evolution by which living things adapt and evolve into different forms over long periods of time through mutations narrowed by natural selection, a mechanism which weeds out those possessing traits less suitable to the environment. The implications of his theory to man's own origins fueled a bitter controversy with Christian orthodoxy.

where he speculated on how environmental factors can affect the characteristics of a species by forcing them to adapt and then passing on those new traits to future generations. However, his work, along with other Arab works, had largely been forgotten. Darwin applied himself to that field of biological knowledge which relates to the breeding of animals and plants, their congenital variations, and the transmission and perpetuation of those variations. On the basis of his observations, Darwin formulated the laws of variation and heredity. According to these, there is a selection among all the congenital variations of each generation of a species. This selection pressure is applied because more young are born than the natural provision of food will support and so there is a struggle for existence and a survival of the fittest. The constant process of selection either maintains the current form of the species from generation to generation or leads to its modification in order to help it adapt to changes in its habitat and environment.

## DARWIN'S ILLUSTRATION OF EVOLUTION

This drawing made by Darwin in 1889 shows the beaks of four species of finches found in the Galapagos Islands. Darwin drew the conclusion that they all probably came from a common ancestor, but had diversified and evolved to adapt to local food supplies on the different islands.

In his theory on teleology, Darwin supposed that every peculiarity in an organism—of size, shape, color, and so on—must either be of benefit to that organism or have been so to its ancestors. According to the theory of natural selection, structures are present either because they are selected as useful or because they are still inherited from ancestors to whom they were useful, though they are no longer useful to the existing generation.

The foundations of Victorian England were shaken by Darwin's contradiction of the biblical version of creation. However, his theory of evolution was gradually accepted. Darwin's seminal publication *The Origin of Species* convinced most biologists that evolution had occurred, but did not convince them that natural selection was its primary mechanism. For instance, Darwin failed to explain how new species arise. According to him, evolution has to be a slow, gradual process. Darwin himself recognized its limitations and wrote, "Natural

selection acts only by taking advantage of slight successive variations; she can never take a great and sudden leap, but must advance by short and sure, though slow steps." Thus, Darwin conceded that, "If it could be demonstrated that any complex organ existed, which could not possibly have been formed by numerous, successive, slight modifications, my theory would absolutely break down." Such a complex organ would be known as an "irreducibly complex system." This is one composed of multiple parts, all of which are necessary for the system to function. If even one part is missing, the system will fail to function. Such a system could not have evolved slowly, piece by piece. Today Darwin's theory of evolution stands challenged by the tremendous advances made in molecular biology, biochemistry, and genetics. It is now known that there are in fact tens of thousands of "irreducibly complex systems" at the cellular level. Even the smallest bacteria are made up of intricate molecular machinery, capable of complex interactive functions, which do not fit into Darwin's model.

## CHARLES DARWIN

Darwin explained life as a parade of mutations evolving through a process called natural selection.

# GENETICS

Experimenting with the pea plant, Gregor Mendel (1822–1884) first traced inheritance patterns of certain traits and showed that they obeyed simple statistical rules. Mendel defined the concept of a fundamental unit of heredity that he called an allele. The term allele as Mendel used it is nearly synonymous with the term gene, and now means a specific variant of a particular gene. Mendel's work was published in 1866 as *Versuche über Pflanzen-Hybriden* (*Experiments on Plant Hybridization*). This work remained unknown for almost 50 years till 1900 when Hugo de Vries, Carl Correns, and Erich von Tschermak independently arrived at the same conclusion as Mendel.

In cross-pollinating plants producing either yellow or green seeds, Mendel showed that the first offspring generation (f1) always has yellow seeds, but the next generation (f2) always has a 3:1 ratio of yellow to green seeds. Mendel's work led him to three important conclusions about inheritance:

• The inheritance of each trait is determined by "units" or "factors" that are passed on to descendents unchanged (these units are now called genes).

• An individual inherits one unit from each parent for each trait.

• A trait may not show up in an individual but can still be passed on to the next generation.

While Mendel's research was with plants, the basic underlying principles of heredity that he discovered apply to all complex life forms. From the selective cross-breeding of the pea plants over many generations, Mendel discovered that certain traits show up in offspring without any blending of parent characteristics. For instance, the pea flowers are either pink or white. Intermediate colors do not appear in the offspring. Mendel observed seven traits that are easily recognized and apparently only occur in one of two forms:

1. flower color (pink or white)
2. flower position (axial or terminal)
3. stem length (long or short)
4. seed shape (round or wrinkled)
5. seed color (yellow or green)

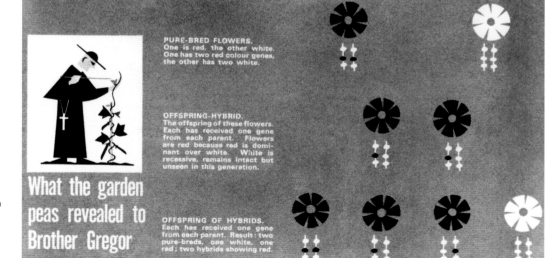

**DRAWING OF MENDEL'S GARDEN PEA STUDIES**

In the 1860s, Gregor Mendel studied the patterns of inheritance as seen in the common, garden pea plant. Mendel's studies helped form the foundation of modern genetics.

6. pod shape (full or constricted)

7. pod color (yellow or green)

In all the seven traits, one form appeared dominant over the other, which is to say, it masked the presence of the other allele. For example, when the genotype for pea seed color is YG (heterozygous), the phenotype is yellow. However, the dominant yellow allele does not alter the recessive green one in any way. Both alleles can be passed on to the next generation unchanged. From his experiments, Mendel derived two fundamental principles of heredity, which later came to be known as the law of segregation and the law of independent assortment. According to the law of segregation, during reproduction, every pair of alleles separate and only one allele of each pair passes from each parent to the offspring. Which allele in a parent's pair of alleles is inherited is a matter of chance. According to the law of independent assortment, different alleles are passed on to offspring independently of each other. This can result in new combinations of genes not present in either parent. For example, a pea plant's inheritance of pink flowers instead of white ones does not affect whether it will have yellow seeds or green ones. In the same way, the law of independent assortment explains why the human inheritance

of a particular eye color does not increase or decrease the likelihood cf having curly or straight hair. Today, we know this is due to the fact that the genes for independently assorted traits are located on different chromosomes. In his time, Mendel did not realize that there are exceptions to these rules. These were discovered later.

## MOLECULAR BIOLOGY

Molecular biology studies the structure and function of cells at the molecular level, chiefly the molecular nature of the gene and its functions like gene replication, mutation, and gene expression. Because of the large space it occupies in the contemporary life sciences, molecular biology would appear to be a much older science. It is in fact a relatively young discipline, originating in the 1930s and 1940s, and becoming institutionalized just two decades later.

The field of molecular biology developed out of the work of geneticists, physicists, and structural chemists on a common problem: the structure and function of the gene. In the early 20th century, genetics was guided by Mendel's two principal laws: the law of segregation (two alleles of a gene segregate during

**THE LAW OF SEGREGATION**

Here violet and white flowered parents produce a first generation (f1) with violet flowers. Self-pollinating these gives a 3:1 ratio of violet to white flowers (f2). Each parent has a pair of alleles controlling flower color. The allele for violet flowers is dominant and "masks" all the white alleles in f1. In f2, 1/4 of the plants have two white alleles.

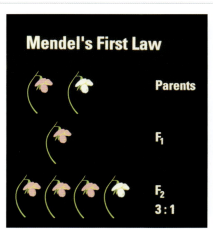

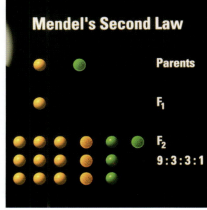

**THE LAW OF INDEPENDENT ASSORTMENT**

Round yellow seeded and wrinkled green seeded parents produce a first generation (f1) with round yellow seeds. Self-pollinating these produces a 9:3:3:1 ratio of round yellow, wrinkled yellow, round green, and wrinkled green seeds (f2). Alleles for yellow seeds and smooth seeds are dominant.

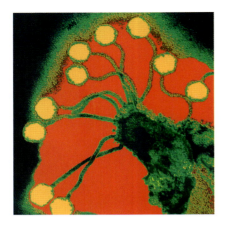

## BACTERIOPHAGE ATTACKING BACTERIA

Hershey and Chase separated the protein coating from the nuclei of bacteriophages, the viruses that infect bacteria. Injecting nucleic acid into the bacterial cell, they found that it was the acid itself, and not the protein, that caused the transmission of genetic information.

## THE PHAGE GROUP

Luria (far left), Delbrück (left) and Hershey shared the 1969 Nobel Prize in physiology for their discoveries concerning the replication mechanism of viruses and their genetic structure.

bacteria infecting virus called bacteriophage. The establishment of the Phage Group in the early 1940s by Delbrueck and Salvador Luria (1912–1991) gave a fillip to molecular biology. In a classical phage experiment in 1952, Alfred Hershey and Martha Chase tracked the chemical components of bacteriophage as they entered bacteria. The results provided evidence, adding to that of Oswald Avery's (1877–1955) earlier work on bacteria that genes were not proteins but deoxyribonucleic acid (DNA). Delbrück's colleague at Cal Tech, Linus Pauling (1901–1991), utilized his knowledge of structural chemistry to study the macromolecular structure of proteins and nucleic acids.

The classical period of molecular biology began in 1953 with the discovery of the DNA double helix by James Watson (1928–) and Francis Crick (1916–2004). They were greatly helped in their work by the X-ray crystallography work on DNA by Maurice Wilkins and Rosalind Franklin, and the model building techniques pioneered by Linus Pauling. Once the structure of DNA was understood, molecular biology shifted its focus to understanding the mechanisms of genetic replication and function, in order to understand the role of genes in heredity. Subsequent research was guided by the notion that the gene was an informational molecule. The linear sequence of nucleic acid bases along a strand of DNA were shown to provide coded information for directing the linear ordering of amino acids in proteins. The genetic code came to be characterized as the relation between a set of three bases on the DNA (a codon) and one of 20 amino acids, the building blocks of proteins.

In 1961, Marshall Nirenberg (1927–) and J. Heinrich Matthaei (1929–) discovered that a unique sequence of nucleic acid bases could be read to produce a unique amino acid product, leading to the later enunciation of the Central Dogma that one gene generates one protein. This "colinearity" between the gene and its function was challenged by two discoveries

the formation of the germ cells), and the law of independent assortment (genes assort independently in the formation of germ cells). The finer mechanisms of gene reproduction, mutation, and expression were not known at that time. Thomas Hunt Morgan (1866–1945) and his colleagues utilized the fruit fly, *Drosophila*, as a model organism to study the relationship between the gene and the chromosomes in the hereditary process. Later, Hermann J. Muller (1890–1967), Morgan's student, recognized the "gene as a basis of life," and began to investigate its structure. Muller discovered the mutagenic effect of X-rays on *Drosophila*, and utilized this phenomenon as a tool to explore the size and nature of the gene. In the next few years physicists began to examine the biological basis of inheritance.

In 1930, the physicist Max Delbrück (1906–1981) began the study of a unique characteristic of life: self-reproduction, using the

in the late 1970s: the overlapping genes and the "split genes." In the case of overlapping genes, two different amino acid chains might be read from the same stretch of nucleic acids by starting from different points on the DNA sequence. In the case of split genes, stretches of DNA often split between coding regions (exons) and non-coding regions (introns). The exons might be separated by large tracts of non-coding, supposedly "junk DNA." The distinction between exons and introns became even more complicated when alternative splicing was discovered in 1978. The phenomenon of alternative splicing enables a series of exons to be spliced together in a variety of ways, thus generating a whole new range of molecular products than the Central Dogma can account for. Discoveries such as overlapping genes, split genes, and alternative splicing forced molecular biologists to rethink their understanding of what actually made a gene and stood the Central Dogma on its head.

Also in the 1970s, molecular biologists developed a variety of techniques for manipulating the genetic material. The recombining of DNA from different species was made possible by the discovery of restriction enzymes that "cut" DNA at specific sites and ligases that "pasted" these DNA segments together. A segment of DNA from one species could be removed and spliced into the DNA of another species. Such species were called "transgenic" forms and were used in practical applications, causing concern about hazards to the environment, and human and animal health. This concern about hazards has grown over time as scientific research has so far failed to bring clarity about the exact nature and extent of biosafety concerns generated by transgenics.

François Jacob (1920–), Jacques Monod (1910–1976) and their colleagues at the Institute Pasteur in Paris discovered that gene induction and regulation were coordinately controlled in Escherichia coli. The group of coordinately controlled genes and their regulatory DNA sites was called an "operon." This discovery of coordinated gene regulation provides a general, theoretical model for transforming descriptive embryology into molecular developmental biology. As molecular biology grew, it branched off into different fields like molecular cell biology, molecular evolution, and molecular medicine.

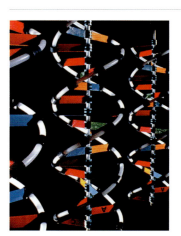

**WATSON AND CRICK (RIGHT) AND DNA DOUBLE HELIX MODEL (LEFT)**

The discovery of the double helix, the twisted-ladder structure of DNA, by James Watson and Francis Crick (far right) in 1953 marked a milestone in the history of science and gave rise to modern molecular biology, which is largely concerned with understanding how genes control the chemical processes within cells. Their discovery led to ground-breaking insights into the genetic code and protein synthesis.

## The Genomic Level

The next stage in the development of molecular biology was its entry into the genomic level. With the knowledge that genes did not work in isolation and often interacted with each other in a phenomenon called epistasis, and with many other components of the cell, it became necessary to accommodate a more sophisticated picture of genes within the larger genome. The genome of an organism is a collection of nucleic acid base pairs: adenine (A) pairs with thymine (T) and cytosine (C) with guanine (G)). The number of base pairs varies widely among species. For example, Haemophilus influenzae has roughly 1.8 million base pairs in its genome while humans have more than 3 billion base pairs.

A technological development which has become fundamental to research on molecular genetics is the Polymerase Chain Reaction (PCR), a procedure by which small samples of DNA can be multiplied to give working samples

## The Human Genome Project

In the mid 1980s, after the development of sequencing techniques, the United States Department of Energy (DoE) began a project to sequence the human genome. This was begun initially as part of a larger plan to determine the impact of radiation on the human genome induced by the Hiroshima and Nagasaki bombings. The resulting Human Genome Project (HGP) utilized both existent sequencing methodologies and introduced new ones. As the project progressed, a controversy emerged between the public Consortium of international sequencing centers and the private sequencing corporation Celera in their race to generate a "rough draft" of the human genome. As it was, Craig Venter's (1946–) Celera beat the multi-agency Human Genome Project to first decipher the human genome in 2000. We now know that humans

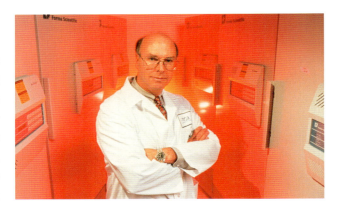

**J. CRAIG VENTER**

Founder of The Institute for Genomic Research (TIGR) in Gaithersburg, Maryland, USA. He is standing beside cryostorage units at TIGR. These units freeze samples of human cDNA (complementary DNA) to a temperature of −80 Celsius for later use in gene research. TIGR was founded in 1992 as the largest gene sequencing laboratory in the world. Venter is currently president of the J Craig Venter Institute, which was created by TIGR's board in 2006.

and most mammals like mice, cats, rabbits, monkeys, and apes have roughly the same number of nucleotides in their genomes—about 3 billion base pairs. Gene duplication occurs frequently in complex genomes; sometimes the duplicated copies degenerate to the point where they no longer are capable of encoding a protein. However, many duplicated genes remain active and over time may change enough to perform a new function. Since gene duplication is an ongoing process, mice may have active duplicates that humans do not possess, and vice versa.

However, the most significant differences between mice and humans are not in the number of genes, which is very similar, but in the structure of genes and the activities of their protein products. Important are the subtle changes accumulated in each of the approximately 25,000 genes, which add up to make quite different organisms. In addition, genes and proteins interact in complex ways that multiply the functions of each. A gene can produce more than one protein product through alternative splicing or post-translational modification; these events do not always occur in an identical way in the two species. A gene can produce more or less protein in different cells at various times in response to developmental or environmental cues, and many proteins can express disparate functions in various biological contexts. Thus, subtle distinctions are multiplied by the more than 30,000 estimated genes.

Similarities between mouse and human genes range from about 70 per cent to 90 per cent, with an average of 85 per cent, but there's a lot of variation from gene to gene (so that some mouse and human gene products are almost identical, while others are completely unrelated). Some nucleotide changes are "neutral" and do not yield a significantly altered protein. Others would introduce changes that could substantially alter what the protein does. Seen in the context of inherited human diseases, a single nucleotide change can mean the difference between health and debilitating disease. Such changes can lead to sickle cell disease, cystic fibrosis, or breast cancer. Single nucleotide changes have been linked to hereditary differences in height, brain development, facial structure, pigmentation, and many other striking morphological differences. Yet, single-nucleotide changes in the same genes but in different positions in the coding sequence may have no harmful impact.

In addition to the human genome, the genomes of about 800 organisms have been sequenced in recent years. These include the mouse *Mus musculus*, the fruitfly *Drosophila melanogaster*, the worm *Caenorhabditis elegans*, the bacterium *Escherichia coli*, the yeast *Saccharomyces cerevisiae*, the plant *Arabidopsis thaliana*, and many microbes.

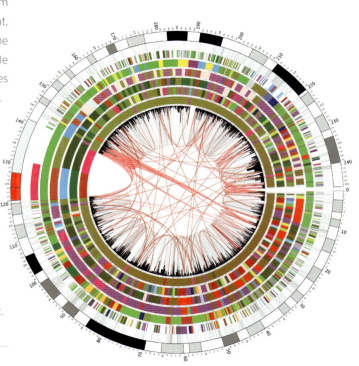

## THE COMPARATIVE GENOME SIZES OF HUMANS AND OTHER ORGANISMS

| ORGANISM | ESTIMATED SIZE (BASE PAIRS) | ESTIMATED GENE NUMBER | AVERAGE GENE DENSITY | CHROMOSOME NUMBER |
|---|---|---|---|---|
| Human | 3.2 billion | 25,000 | 1 gene per 100,000 bases | 46 |
| Mouse | 2.6 billion | 25,000 | 1 gene per 100,000 bases | 40 |
| Fruit fly | 137 million | 13,000 | 1 gene per 9,000 bases | 8 |
| Plant | 100 million | 25,000 | 1 gene per 4,000 bases | 10 |
| Roundworm | 97 million | 19,000 | 1 gene per 5,000 bases | 12 |
| Yeast | 12.1 million | 6,000 | 1 gene per 2,000 bases | 32 |
| Bacteria | 4.6 million | 3,200 | 1 gene per 1,400 bases | 1 |

### CIRCULAR GENOME MAP

Map showing shared genetic material between humans (outer ring) and (from inner ring outward) chimpanzee, mouse, rat, dog, chicken, and zebrafish chromosomes. Each ring is based on the idiogram of one chromosome. The colors within each ring form a heat map, the pattern of which represents hot spots of shared genetic material, or synteny. A highly fragmented pattern indicates greater evolutionary divergence from humans. The numbers on the human chromosome represent scale bars and the gray intersecting lines are paths indicating chromosome loci.

# Genomics

Genomics is a new field of study that characterizes the complete genetic makeup of an organism. Its approach is to determine the entire sequence and structure of the organism's DNA, which is its genome, and then to determine how the DNA in the genome is arranged into genes. This second part is accomplished by determining the structure and relative abundance of all messenger RNAs (mRNAs), the middlemen in genetics that encode individual proteins.

In the initial phase, the study of genomics was directed at microorganisms, which have relatively small genomes. More recently, because of more industrialized, higher-throughput sequencing technologies, the genomes of many organisms have been sequenced. Genomics has the potential to provide great breakthroughs in controlling infectious diseases,

leading to the development of new disease prevention and treatment strategies for plants, animals, and humans. For instance, understanding pathogen genes, their expression, and their interaction will lead to new antibiotics, antiviral agents, and "designer" immunizations. These new DNA-based immunizations are the by-products of genomic research and will perhaps eventually replace the traditional vaccines made from whole, inactivated microorganisms.

Understanding the genomes of plants and animals has additional benefits. Gene mapping should allow us to understand the basis for disease resistance, disease susceptibility, weight gain, and determinants of nutritional value. The use of genomic information provides the opportunity to select optimal environments for the healthy growth of plants and animals, to develop disease-resistant strains, and to achieve improved nutritional value. Bacteria,

**GENOMICS**

Conceptual computer artwork of an anatomical model of the human face with a DNA helix (across bottom) and three molecular models (upper right). This may represent genomics, the process of determining the sequences of genes in the human genome (the information that controls and transmits an organism's hereditary characters). The sequence of DNA's nucleotide bases (G, C, A, T), arranged in discrete segments known as genes, determines the genetic code. Each gene encodes instructions for the construction of molecules like proteins. Sequencing the genome allows identification of the genes responsible for genetic diseases.

viruses and fungi play important roles in agriculture. Understanding the genomics of these organisms has the potential to improve crop yields. decrease damage done by pest species, and increase the nutritional value of food. As part of their metabolism, some microorganisms have the ability to break down harmful products and to produce energy as a product. Understanding the gene products involved in these transformations may lead to industrial uses, with the potential for solving different types of environmental problems and providing new energy sources.

Understanding the function of genes and other parts of the genome is known as functional genomics. The Human Genome Project was just the first step in understanding humans at the molecular level. Though the project is complete, many questions still remain unanswered, including the function of most of the estimated 30,000 human genes. Researchers still don't know the role of single nucleotide polymorphisms (SNPs)—single DNA base changes within the genome—or the role of noncoding regions and repeats in the genome. Functional genomics research is conducted using model organisms, such as mice, to follow the inheritance of genes that are very similar to human genes.

Comparative genomics is the analysis and comparison of genomes from different species to gain a better understanding of how species have evolved and to determine the function of genes and noncoding regions of the genome. Scientists have learned about the function of human genes by examining their counterparts in simpler model organisms such as the mouse. Genome researchers look at many different features when comparing genomes: sequence similarity, gene location, the length and number of coding regions (called exons) within genes, the amount of noncoding DNA in each genome, and highly conserved regions maintained in organisms ranging from bacteria to humans. Comparative genomics involves the

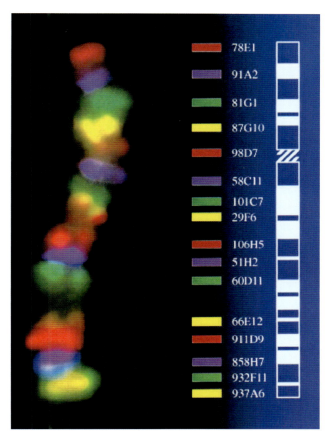

**CHROMOSOMAL RAINBOW**

Fluorescent light micrograph and diagram of probes bound to human chromosome 10. Chromosomes are composed of DNA coiled around proteins. DNA contains sections, called genes, which encode the body's genetic information. In this technique, each block of color is a probe that binds to a specific DNA sequence. During meiosis (the formation of the sex cells), pairs of chromosomes line up together and may overlap and swap segments (recombination). This technique allows the rapid mapping of these chromosomal rearrangements as well as abnormalities such as deletions and insertions of blocks of DNA.

use of computer programs that can arrange multiple genomes to identify regions of similarity.

Structural genomics is the systematic study of the three-dimensional structures of proteins that represent the range of

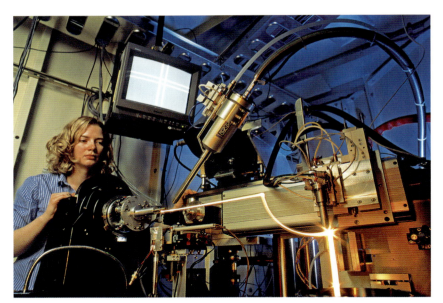

**PROTEIN CRYSTALLOGRAPHY**

A researcher is seen here with an automatic X-ray crystallography machine used to study the structure of proteins. The light track shows the movement of a robot arm as it selects a sample from lower right and places it in position for crystallography at center left. Crystallography is used to determine the three-dimensional structure of a molecule. The diffraction pattern produced as X-rays pass through a crystal, reveals its structure. This work is being done at Lawrence Berkeley Laboratories, California, USA. This automated protein crystallization and crystallography process is much faster than conventional non-automated methods.

protein structure and function found in nature. The ultimate aim is to build a body of structural information that will facilitate prediction of a reasonable structure and potential function for almost any protein, from knowledge of its coding sequence. Such information will contribute to understanding the functioning of the human proteome, consisting of thousands of proteins. Structural genomics is a new and rapidly growing inter-disciplinary research, aiming to generate a systematic database of protein structures. Synchrotron radiation-based X-ray crystallography is unique in providing very accurate high resolution structures of proteins and their complexes. Translating genome sequence to corresponding protein structures via high throughput approaches has great significance for medical diagnostics and healthcare as well as for fundamental biology.

The progress in structural genomics has been enabled by the development of several related key technologies. These include synchrotron-based MAD (Multiple wavelength Anomalous Dispersion), phase determination, cloning and recombinant expression, genome sequencing, and bioinformatics.

## METABOLOMICS

Metabolomics is the study of the naturally occurring molecules, called metabolites, in biological materials such as cells, biofluids, and tissues. These small molecules (1500 daltons or less) are the products of metabolism and include, for example, sugars, fats, and amino acids. The sum of all the metabolites within a cell is called the metabolome. The study of metabolomics requires inputs from various disciplines like chemistry, biology, physics, computer science, and statistics,

The three principal approaches to studying the metabolome are metabolite profiling, metabolic fingerprinting, and metabonomics. Metabolite profiling aims to identify and quantify metabolites. Despite its methodological limitations, this technique can, if done accurately, give a true and complete

## NMR SPECTROMETRY

A scientist at the controls of an NMR spectrometer. This measures the resonance between an applied oscillating magnetic field and the magnetic moment of atoms. The display seen here is a "2D" trace. This is a plot of excitation frequency on one axis, resonance frequency on the second axis, and response intensity in the vertical axis. This type of plot is used to identify the constituents of a compound and to give some data relating to its structure.

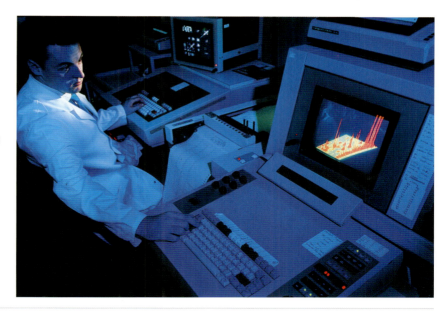

visualization of the metabolome. Metabolic fingerprinting is a high-throughput approach utilized in tissue comparison or discrimination analysis. Metabonomics focuses on the metabolic response of organisms to pathophysiological stimuli or genetic modification. This approach is generally restricted to microbiological and non-botanical studies.

The tools and technologies used to measure the metabolites include gas chromatography-mass spectrometry (GC/MS), nuclear magnetic resonance (NMR) spectroscopy, mass spectrometry for exact mass determination, capillary electrophoresis systems, nano-HPLC systems for rapid compound separation, and fourier transform ion cyclotron resonance (FTICR). Metabolomics research is opening up new horizons and being applied to different fields, from plant science to medicine. In plant science, metabolomic research is examining biomass accumulation, stress resistance, and secondary metabolite production. In medicine, metabolomics helps to detect the presence and severity of coronary heart disease. It also helps in the prediction of the clinical outcome of a sudden hemorrhage of a blood vessel. In the field of environment, researchers are applying these methods to study the effects of both diseases and chemicals on wildlife species. Environmental metabolomics may be useful in measuring the interactions of genes with the environment. The advantage of metabolomics for disease diagnosis, whether in humans or animals, stems from the fact that when an organism becomes diseased or stressed, molecular/phenotypic changes are observed in the body. These changes can be measured using metabolomics.

## PROTEOMICS

The term proteome was first suggested in 1994 by Marc Wilkins. It refers to proteins that are encoded and expressed by a genome. Proteomics can be defined as the systematic large-scale analysis of protein expression under normal and stressed states, and generally involves the separation, identification, and characterization of all of the proteins in a cell or tissue sample. The key methods used in proteomics are 2-D gel electrophoresis (2D-GE) and mass spectrometry (MS). Gel electrophoresis is used to separate the proteins by size and isoelectric part. The individual proteins are subsequently removed from the gel and analyzed by MS to determine their identity and characteristics.

Proteomics technology is being used for cancer diagnosis and treatment. It primarily identifies biomarkers—proteins and protein patterns in blood, urine, and tissue that can be used to detect the likelihood of early cancers and treatment response. Hence proteomics is expected to improve cancer treatments, predict the effects of various treatments and develop individualized therapies for each patient. Proteomics has led to the identification of many biomarker proteins and the discovery of many new proteins in the blood. It has been used to identify hundreds of proteins in the ovary, prostate, breast, and esophagus that increase or decrease as cells begin to grow abnormally. Proteomics researchers are concentrating on ovarian and prostate cancers, which usually do not get detected in early stages. By using proteomics for early detection, tumors may be treated before they spread to other parts of the body.

## STEM CELLS

A stem cell is essentially the building block of the human body. The stem cells inside an embryo will eventually give rise to every cell, organ, and tissue in the fetus's body. Unlike a regular cell, which can only replicate to create more of its own kind of cell, a stem cell is pluripotent. When it divides, it can make any one of the 220 different cells in the human body. Stem cells also have the capability to self-renew—they can reproduce themselves many times over. Stem cells have different differentiation potential, referred to as potency. Totipotent stem cells arise from the fusion of egg and sperm cell. Pluripotent stem cells

**PROTEOMICS RESEARCH**

Proteomics is the study of the structure and function of all the proteins in an organism. Here a technician is studying a 2-D gel of heart proteins. The gel separates proteins by their charge and mass by the application of a pH gradient and an electric field respectively. Each spot here contains similar proteins, which can be removed from the gel and analysed. Proteins form nearly all the functional tissues in an organism. Finding how structure affects function will improve understanding of how organisms work and may lead to many new targets for drugs.

are close descendents of totipotent stem cells. They have wide differentiation powers, but are limited to cell types of the three germ layers—ectoderm, endoderm, and mesoderm.

There are two types of stem cells—embryonic stem cells and adult stem cells. Embryonic stem cells come from an embryo—the mass of cells in the earliest stage of human development that, if implanted in a woman's womb, will eventually grow into a fetus. Adults also have stem cells in the heart, brain, bone marrow, lungs, and other organs. They are our built-in repair kits, regenerating cells damaged by disease, injury, and everyday wear and tear. The earlier belief that adult stem cells were to be found in limited numbers in the body, giving rise to the same type of tissue from which they originated, has been disproved by new researches. Currently, stem cells have been harvested from a diverse range of tissues like cord blood and dental tissue. These suggest that adult stem cells may have the potential to generate other types of cells as well. For example, liver cells may be coaxed to produce insulin, which is normally made by the pancreas. This capability is known as plasticity or transdifferentiation.

Stem cells may help us understand how a complex organism develops from a fertilized egg. Scientists can follow stem cells as they divide and become increasingly specialized, making skin, bone, brain, and other cell types. Identifying the signals and mechanisms that determine whether a stem cell chooses to carry on replicating itself or differentiate into a specialized cell type, and into which cell type, will help us understand what controls normal development. An important goal of stem cell research is to gain a better understanding of diseases, how they arise, and suggest new strategies for therapy. Stem cells may hold the key to replacing cells lost in devastating diseases such as Parkinson's, stroke, heart disease, and diabetes, for which there are currently no sustainable cures.

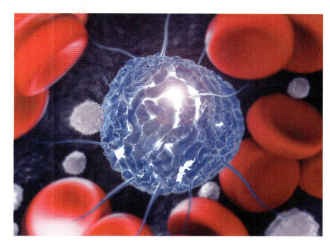

**CONCEPTUAL ARTWORK OF STEM CELL AND RED BLOOD CELLS**

Stem cells can differentiate into any other cell type. There are three main types of mammalian stem cell: embryonic stem cells, derived from blastocysts; adult stem cells, which are found in some adult tissues; and cord blood stem cells, which are found in the umbilical cord. The cells seen here are destined to become blood cells. During blood cell development in adults, stem cells develop through a process known as haemopoiesis.

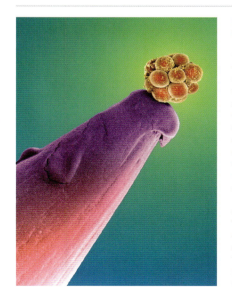

**EMBRYONIC STEM CELLS**

Coloured scanning electron micrograph (SEM) of a human embryo at the 16-cell stage on the tip of a pin. The ball of cells (yellow) of the embryo is known as a morula, a cluster of almost identical, rounded cells, each containing a central nucleus. This 16-cell embryo is about three days old. It is at the early stage of transformation from a single cell to a human composed of millions of cells.

Stem cells can be obtained from several sources. They can be obtained from spare embryos stored at fertility clinics or from special purpose embryos created in vitro fertilization (artificially in the lab) for the sole purpose of extracting their stem cells, or from embryos cloned in labs using somatic nuclear transfer method in order to harvest their stem cells. They are also taken from fetuses in early development that have been aborted, from umbilical cords (which hold great potential for research), and from the tissue or organs of living adults during surgery. The isolation and survival of neural progenitor cells from human post-mortem tissues (up to 20 hours after death) has been reported and provides an additional source of human stem cells.

Another major application of stem cell research is its use in therapies. Stem cells can renew blood and bones after chemotherapy. Bone marrow transplants (BMT) are a well known clinical application of stem cell transplantation. BMT can repopulate the marrow and restore all the different cell types of the blood after high doses of chemotherapy and/or radiotherapy, the main defense used to eliminate endogenous cancer cells.

The knowledge of stem cells has made it possible for scientists to grow skin from a patient's plucked hair. Skin (keratinocyte) stem cells reside in the hair follicle and can be removed when a hair is plucked. It is presently being studied in clinical trials as an alternative to surgical grafts used for ulcers and burn victims. Stem cells can provide dopamine—a chemical lacking in victims of Parkinson's Disease. Parkinson's Disease involves the loss of cells which produce the neurotransmitter dopamine. The first double-blind study of fetal cell transplants for Parkinson's Disease reported survival and release of dopamine from the transplanted cells and an improvement of clinical symptoms.

Mouse stem cells were made to produce their own insulin. Recently, insulin secreting cells from mouse stem cells have been generated. In addition, the cells self assemble to form structures that closely resemble normal pancreatic islets and produce insulin. Future research will need to investigate how to optimize conditions for insulin production with the aim of providing a stem cell-based therapy to treat diabetes to replace the constant need for insulin injections.

## SYSTEMS BIOLOGY

Systems biology is the integrated study of the network of genes, proteins, and biochemical reactions which make up a living organism. No biological system is the result of a single mechanism or gene. Brain functions or the immune systems are the result of the interactions of numerous genes, proteins, mechanisms, and stimuli from the external environment.

Systems biology attempts to understand these interactions. The individual function and collective interaction of genes, proteins, and other components in an organism are often characterized together as an interaction network. Understanding this interaction network and the interplay of an organism´s genome and external environmental influences is crucial to developing a systems understanding of an organism that will ultimately transform our understanding of human health and disease. Traditional biology focuses on identifying individual genes, proteins, and cells, and studying their specific functions. Such an approach in biology doesn't enable us to fully understand how, for instance, the human body operates, and how we can best predict, prevent, or remedy potential health problems. The systems biology approach requires determining the effect of every gene or cluster of genes and understanding how that works as a system.

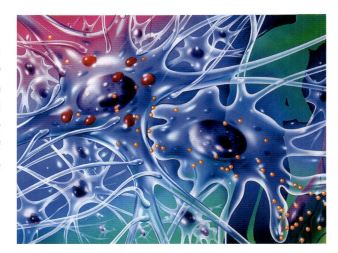

**PARKINSON'S DISEASE TREATMENT**

Artwork of Parkinson's disease nerve cells (neurones) being treated with the drug Levodopa (L-dopa). The neurones, with dark blue nuclei and long nerve processes, are in the basal ganglia nerve clusters in the center of the brain. In the neurones are Lewy bodies (red) which are believed to cause them to degenerate. Here, L-dopa (L-dihydroxyphenylalanine) is being converted into dopamine (orange balls) as it moves into the brain from a blood vessel (bottom right).

**GENETIC ANALYSIS**,
Photographed in 2003, at the
Centre de Bioingenierie Gilbert
Durand, Toulouse, France

Fluorescent samples of DNA
(colored spots) on a glass slide,
forming a DNA microarray (also
called a DNA chip or gene array).
DNA is the genetic cellular
material that is a fundamental
basis of the biochemistry of life.
The microarray can be used to
investigate and compare the
properties of thousands of genes
at once, a process known as
expression profiling.

Unlike genetic engineering (transfer of genes across the species barrier) and synthetic biology (where few genes are used to construct a life form), systems biology examines the components of a system together, rather than taking them apart. It moves beyond the one gene-one protein approach and focuses on a detailed understanding of how large numbers of interrelated parts of a living system make up networks whose functional properties are reflected in subtle differences in expression, thus resulting in different individuals with different phenotypes. Gathering information at all levels and examining complex interactions among molecules, as systems biology does, has the potential to help in designing treatments more precisely. Scientists are studying gene expression, protein profiling, and metabolite profiling all at the same time, as related processes, to understand a particular phenotype, for instance a particular disease. The greatest promise of systems biology today appears to be in the medical field where an attempt is being made in the area of predictive, preventive, and personalized medicine. However, the subject is still in a nascent stage. Its experimental, technical, computational, and sociological challenges are still largely unresolved. A major challenge seems to be getting biologists, engineers, computer scientists, and mathematicians to agree on the choice of a model system and biological process, and the strategies for investigating it.

Scientists hope that systems biology research, which includes model building, hypothesis formation, data collection, and analysis, would result in accurate quantitative predictions of the behavior of genes and proteins in times to come. This would help in gaining an insight into the basic biological processes of cells. Though these modules and networks may be quite complex, involving hundreds of genes and proteins, insights gained from yeast and other simple model systems are already likely to assist with the practical design of delivery of pharmaceuticals.

# SYNTHETIC BIOLOGY

Synthetic biology is one of the new biologies. Not really biology in the classical sense, it is a mixture of engineering, biology, chemistry, and physics, which tries to reconstruct life at the genetic level. Synthetic biology refers to the design and fabrication of biological components and systems that do not exist in the natural world. It also attempts to re-design existing biological systems to build those that behave in predictable ways. The era of synthetic biology has been described as an era in which significantly new gene arrangements can be constructed and evaluated for a diverse range of applications.

Craig Venter, the scientist who raced ahead of the multi government Human Genome Project in 2000 to crack the human genome first, is also in the forefront of the most recent breakthrough in synthetic biology. Venter's group created a new organism, which they have termed *Mycoplasma laboratorium*. Starting with the bacterium *Mycoplasma genitalium*, the Venter group stripped the bacterium of most of its original DNA and built up a new genome using pieces of DNA synthesized in the laboratory. Most of the work on synthetic biology is happening in the US, with scientists at the Craig Venter Institute in Maryland taking the lead. There are other research groups also working in Europe, Japan, and Israel. Synthetic biology is likely to expand as a discipline as more labs take up this research, especially those where genetic engineering is already established. Although similar in many ways, there is a fundamental difference between genetic engineering which shifts individual (natural) genes from one species to another and synthetic biology which seeks to assemble new genes from bits of DNA that can be synthesized in the lab.

The promises of synthetic biology are many but so are its many potential risks. As a field, it is projected to go far ahead of genetic engineering and perform many functions which are almost unimaginable today. Artificially constructed systems made through synthetic biology, even more precise than Venter's new bacterium, are being projected to produce cheap drugs for malaria and medically useful chemicals. Other applications are the production of industrial chemicals, cellulosic ethanol for use as a biofuel and microorganisms equipped with artificial biochemical pathways aimed to attack specific pollutants and detoxify the environment. Along with the promise of bioremediation, synthetic biology offers more far-fetched promises like making bacterial films that can distinguish light and shade such that visual patterns can be recorded, almost like photographs. This property could have value in diagnostic tools in medicine, for detecting mineral deposits and for mapping the earth's surface from close up.

Despite its progress, quite a few fundamental bottlenecks remain unresolved. The behavior of bioengineered systems

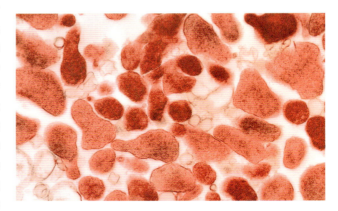

**MYCOPLASMA GENITALIUM**

M. genitalium has the smallest genome (total genetic material) of all living organisms. Its genome was mapped in 1993, making it the second complete bacterial genome to be sequenced. In January 2008, a team at the J. Craig Venter Institute, USA, used this map to make a synthetic bacterial chromosome, called M. laboratorium, from scratch. The new chromosome, containing only the genes needed for life, was then inserted into a M. genitalium bacterium with its genome removed. Having a different chromosome changes the bacterium's function, making it different from wild M. genitalium bacteria.

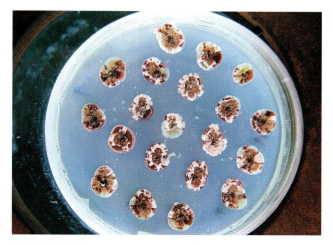

**BIOFUEL BACTERIA**

Petri dish containing colonies of recombinant (genetically modified) Streptomyces bacteria (red) that may produce cellulase. This is the enzyme that breaks down cellulose (the primary structural material in plants). Bacteria that can produce cellulase, or similar enzymes, are able to ferment plant cellulose to produce ethanol for use as a fuel. Fuels produced by biological means (biofuels) are renewable, unlike fossil fuels.

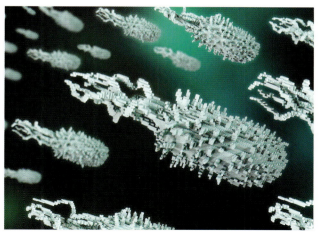

**ENGINEERED BACTERIA**

Bacteria (green) made out of Lego, representing the ability of scientists to alter and rearrange a bacterium's genetic material. Applications of genetically modified (GM) bacteria include the manufacture of synthetic human insulin to treat diabetes and research into the manufacture of certain proteins that can block the Human Immunodeficiency Virus (HIV) from infecting its target cells.

remains unpredictable, because genetic circuits that have been created artificially tend to mutate rapidly and frequently become non-functional. Scientists say that synthetic biology will not deliver unless scientists can accurately predict how a new genetic circuit will behave once it is put into a living cell. This is the same problem that is encountered in genetic engineering. Until scientists are in a position to understand and reliably predict the molecular processes in a cell before and after intervention, the engineering of biological systems either through genetic engineering or through synthetic biology will remain ad hoc and undependable, ultimately even dangerous.

Research in synthetic biology is focused in several areas but perhaps its most promising field is making DNA computers. DNA computing is based on the premise that, like a computer, DNA both stores and processes coded information. DNA computing was born in 1994 when Leonard Adleman demonstrated how to

solve a complex computational problem (the answer to which was known), using DNA to sort through possible answers and find the correct one.

While systems biology studies complex biological systems as integrated wholes focusing on natural systems, often with long term medical interest, synthetic biology studies the building of artificial biological systems for engineering applications, using many of the same tools and experimental techniques. But the work is fundamentally an engineering application of biological science rather than an attempt to do more science. Genetic engineering looks at one gene at a time but synthetic biology tries to build genetic pathways. It involves imbuing biological forms with new functions, creating new forms from existing biological components at the genome level. Bioengineering applies engineering principles to biology, to aid scientific discovery for clinical applications such as tissue transplants or diagnostic machines. Synthetic biology is quite similar to nanobiotechnology. The programming and functioning of living machines in the future will frequently involve the integration of biological and non-biological parts at the nano level. Nanobiotechnology tries to merge the living and non-living systems at the nano scale to make hybrid materials and organisms.

## BIOINFORMATICS

Bioinformatics is the use of computer software and programs to organize, store, and analyse biological information and data to better understand biological systems. It is a field where biology, computer science, and information technology converge as a single discipline. The goal of bioinformatics is to get new biological insights and try to identify unifying principles in biology. Bioinformatics involves the use of advanced computer and statistical techniques to organize and analyse large amounts of biological data generated through modern biotechnologies. Genetic engineering and the "omics" technologies, such as genomics proteomics and metabolomics, are giving rise to very large numbers of scientific information, which must be organized if they are to make any sense.

The field of bioinformatics developed as a corollary to the creation of the GenBank database, which was started by the US Department of Energy in the early 1980s. This database stores DNA sequence information obtained from a number of organisms. When researchers could access data from all over the world, the DNA sequence data in GenBank grew rapidly with the emergence of highly sophisticated gene sequencing tools. Private companies joining the sequencing race with parallel projects created huge databases of their own. The two most significant services enabling access to bioinformatics are the European Molecular Biology Network (EMBnet) and the United States National Center for Biotechnology Information

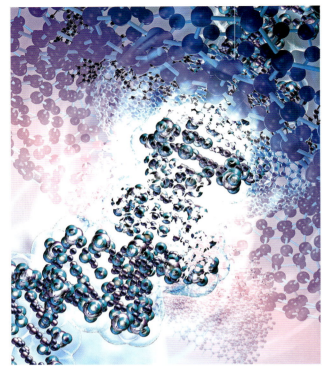

**DNA NANOTECHNOLOGY**

Conceptual computer artwork of DNA nanotechnology, showing carbon nanotubes (allotropes of carbon) at top right and a molecule of DNA at bottom left. DNA nanotechnology is the use of DNA in technologies on the scale of nanometres (billionths of a meter). One example is nanotube structures that resemble their carbon counterparts. DNA nanotubes, self-assembling superstructures composed of DNA tiles, may be used in applications ranging from new structural materials that are stronger and lighter, to electronic components for new supercomputers and drug delivery systems.

(NCBIO). In Canada, the Canada Institute for Scientific and Technical Information (CISTI) of the National Research Council of Canada (NRC) operated a bioinformatics server, Molecular Biology Database Service (MBDS), from 1987 to 1996 after which the Canadian Bioinformatics Resource (CBR) took over. Bioinformatics addresses three main areas: examining relationships between the different sets of biological data; analysing and interpreting the different categories of data,

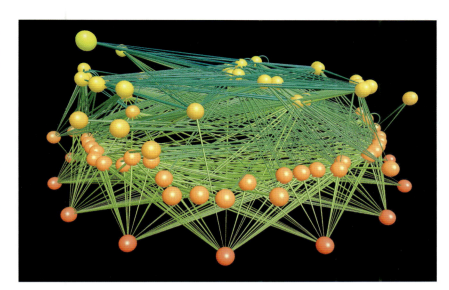

**ECOLOGICAL FOOD WEB,** Model created by Neo Martinez, Pacific Ecoinformatics and Computational Ecology Lab, USA

This model represents the aquatic ecosystem of the East River Valley, Denver, Colorado, USA. Animal and plant life is represented by the colored nodes. Red nodes are lowest trophic level organisms such as plants and detritus, orange represents intermediate species and yellow represents top-level species and primary predators. The connecting lines show relationships between organisms. Thick lines connect to predator species and thin lines connect to prey. Food webs are used by biologists to predict ecological outcomes.

which could include nucleotide and amino acid sequences; creating computer/software tools to enable systematic access, and management of different categories of data.

Using bioinformatic tools, researchers are able to compare the genomes of species to identify similarities and differences among organisms; identify DNA sequences of identified genes from published material on the internet; and identify information about proteins and their amino acid sequences. Once the protein sequences are analysed, researchers can use software programs to search databases to find similar sequences to detect protein homology across organisms. Sequence translation programs enable users to analyse nucleotide sequences and convert the DNA sequences into protein sequences or protein sequences into their complementary DNA. Bioinformatic programs also allow researchers to look at the three-dimensional shape of proteins and, based on this, the functions of the proteins can be determined. A good bioinformatics program should be able to combine the available information to provide a comprehensive picture of normal

cellular activities so that researchers can study how these activities are altered in different disease states.

The process of analyzing and interpreting data is referred to as computational biology, which involves the development of new algorithms and statistics to assess relationships among members of large data sets. These could be methods to locate a gene within a sequence, predict protein structure and function, or cluster protein sequences into families of related sequences. Computational biology uses applied mathematics, informatics, statistics, computer science, artificial intelligence, chemistry, and biochemistry to solve biological problems at the molecular level. Bioinformatics aims to increase our understanding of biological processes. Major research efforts in the field include sequence alignment, gene detection, genome assembly, protein structure alignment, protein structure prediction, prediction of gene expression and protein-protein interactions.

Informatics has assisted evolutionary biologists in several ways. It has enabled researchers to trace the evolution of a large number of organisms by measuring changes in their DNA, rather than through physical taxonomy or physiological observations as was done earlier. It is now possible to compare entire genomes, which permits the study of complex evolutionary events, such as gene duplication and horizontal gene transfer.

## CLONING

One of the most controversial issues in contemporary biology is that of cloning. The term clone is derived from the Greek word for "twig" or "branch", referring to the process whereby a new plant can be created from a twig. Organism cloning refers to the procedure of creating a new multicellular organism, genetically identical to another. In essence, this form of cloning is an asexual method of reproduction.

## REPRODUCTIVE CLONING

Embryos are cloned for use in research, particularly stem cell research. This process is also called "research cloning" or "therapeutic cloning." The goal is not to create cloned human beings, but rather to harvest stem cells that can be used to study human development and to potentially treat disease.

Reproductive cloning uses "somatic cell nuclear transfer" (SCNT) to create animals that are genetically identical. This process entails the transfer of a nucleus from a donor adult cell (somatic cell) to an egg that has no nucleus. If the egg begins to divide normally, it is transferred into the uterus of the surrogate mother. Such clones are not strictly identical since the somatic cells may contain mutations in their nuclear DNA. Additionally, the mitochondria in the cytoplasm also contains DNA and during SCNT this DNA is wholly from the donor egg, thus the mitochondrial genome is not the same as that of the nucleus donor cell from which it was produced. This may have important

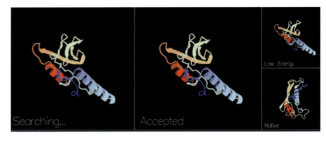

**ROSETTA@HOME SCREEN SAVER**

Computer graphic showing possible protein shape configurations. Rosetta@Home is a volunteer or distributed computing project designed to predict protein folding and shapes, as found in nature. Volunteer computing allows scientists to replicate supercomputing power by utilizing the idle time of personal computers owned by the public. Data is sent to a user via the internet and the screen saver performs calculations when the computer is unused. The results are sent back to a base server. The results will help scientists learn more about diseases and could lead to new drugs and cures. Rosetta@Home runs on the Berkeley Open Infrastructure for Network Computing (BOINC).

implications for cross-species nuclear transfer in which nuclear-mitochondrial incompatibilities may lead to death.

## DOLLY THE SHEEP

Dolly, an ewe, became in 1996 the first mammal to have been successfully cloned from an adult cell, though the first actual animal to be cloned was a tadpole in 1952. She was cloned at the Roslin Institute in Scotland and lived there until her death when she was six. In April 2003, her stuffed remains were placed at Edinburgh's Royal Museum, part of the National Museums of Scotland.

Dolly was publicly significant because the effort showed that the genetic material from a specific adult cell, programmed to express only a distinct subset of its genes, can be reprogrammed to grow an entire new organism. Before this demonstration, there was no proof for the widely spread hypothesis that differentiated animal cells can give rise to entire new organisms.

Cloning Dolly the sheep had a low success rate per fertilized egg; she was born after 277 eggs were used to create 29 embryos, which only produced three lambs at birth, only one of which lived. Notably, although the first clones were frogs, no adult cloned frog has yet been produced from a somatic adult nucleus donor cell.

There were early claims that Dolly the Sheep had pathologies resembling accelerated aging. Scientists speculated that Dolly's death in 2003 was related to the shortening of telomeres, DNA-protein complexes that protect the end of linear chromosomes. However, other researchers, including Ian Wilmut who led the team that successfully cloned Dolly, argue that Dolly's early death due to respiratory infection was unrelated to deficiencies with the cloning process.

## HUMAN CLONING

Human cloning is the creation of a genetically identical copy of an existing or previously existing human. The term is generally used to refer to artificial human cloning. Human cloning is usually divided into two types: therapeutic cloning and reproductive cloning. Therapeutic cloning involves cloning cells from an adult for use in medicine and is an active area of research. Reproductive cloning would involve making cloned human beings and such cloning has not been performed and is illegal in many countries. A third type of cloning called replacement cloning is a theoretical possibility, and would be a combination of therapeutic and reproductive cloning. Replacement cloning would entail the replacement of an extensively damaged, failed, or failing body through cloning followed by whole or partial brain transplant.

### IAN WILMUT AND DOLLY

In 1996, British embryologist Professor Ian Wilmut created "Dolly", the world's first sheep cloned from an adult sheep cell. The research was conducted at the Roslin Institute in Edinburgh, Scotland. The cell nucleus was removed from an egg cell taken from a Scottish Blackface ewe. Next, an adult cell from the udder of a 6-year-old Finn Dorset ewe was cultured and injected into the egg cell. A spark of electricity then fused the udder cell with the egg cytoplasm and stimulated the egg to grow into an embryo in the womb of a surrogate sheep.

The first human hybrid human clone was created in November 1998, by American Cell Technologies. It was destroyed after 12 days, as a normal embryo begins to grow at 14 days. In January 2008, scientist Samuel Wood and Andrew French, Stemagen company's chief scientific officer in California, announced that they had successfully created the first 5 mature human embryos using DNA from skin cells donated by Wood and a colleague. These embryos were destroyed and it is not clear if they would have been capable of further development.

## CLONING EXTINCT AND ENDANGERED SPECIES

The reconstruction of functional DNA from extinct species has, for decades, been a dream of some scientists. The possible implications of this were dramatized in the best-selling novel by Michael Crichton and the Hollywood thriller based on it, *Jurassic Park*. But attempts to extract DNA from frozen mammoth fossils have been unsuccessful, though a joint Russo-Japanese team is currently working toward this goal.

**PROMETEA, THE FIRST CLONED HORSE**

Prometea (foreground) with her identical twin and surrogate mother, Stella Cometa. Prometea was created by taking a skin cell from Stella and fusing it with an empty horse egg. The resulting embryo was cultured for five days before being implanted into Stella's uterus. Prometea was born on May 28, 2003, in a laboratory in Cremona, Italy.

In 2001, a cow named Bessie gave birth to a cloned Asian *gaur*, an endangered species, but the calf died after two days. In 2003, a banteng, a species of wild cattle, was successfully cloned, followed by three African wildcats from a thawed frozen embryo. These successes provided hope that similar techniques (using surrogate mothers of another species) might be used to clone extinct species. Anticipating this possibility, tissue samples from the last bucardo (Pyrenean Ibex) were frozen immediately after it died. Researchers are also considering cloning endangered species such as the giant panda, ocelot, and cheetah. The "Frozen Zoo" at the San Diego Zoo now stores frozen tissue from the world's rarest and most endangered species.

In 2002, geneticists at the Australian Museum announced that they had replicated DNA of the Thylacine (Tasmanian Tiger), extinct about 65 years ago, using polymerase chain reaction. However, in February 2005, the museum announced that it was stopping the project after tests showed the specimens' DNA had been too badly degraded by the (ethanol) preservative. Most recently, in May 2005, it was announced that the Thylacine project would be revived, with new participation from researchers in New South Wales and Victoria.

One of the continuing obstacles in the attempt to clone extinct species is the need for nearly perfect DNA. Cloning from a single specimen could not create a viable breeding population in sexually reproducing animals. Cloning endangered species is also a highly ideological issue. Many conservation biologists and environmentalists vehemently oppose cloning endangered species on the grounds that it would deter donations to help preserve natural habitat and wild animal populations. In a 2006 review, the scientist David Ehrenfeld castigated attempts at cloning endangered species for not addressing any of the issues underlying animal extinction, such as habitat destruction, hunting, and an impoverished gene pool.

# GEOLOGY

# Chapter 1
# THE MEASURE OF THE EARTH

**COPERNICUS'S SYSTEM** That it is the Earth which circles the Sun and not the other way round, was a claim that pitted Copernicus against all of Christendom, but it also inaugurated the Scientific Revolution in Europe. This representation of the Copernican system shows how the Sun illuminates different parts of the globe according to the time of day and the season of the year.

**FACING PAGE:** Map tracing Ferdinand Magellan's voyage around the world, the first circumnavigation of the Earth. The voyage began in Seville, Spain, in 1519 and ended in 1522. Magellan never completed the journey. He was killed in a battle at the Philippines. This map was once owned by King Charles V, ruler of the Holy Roman Empire from 1519, and, as Charles I of Spain, of the Spanish realms from 1516 until his abdication in 1556.

**G**eology is the science of the earth (Greek: geo-earth, logos-study). In Greek mythology, Gaia or Ge was the Earth goddess. Geology deals not only with the landform and other surface features of the Earth, but also with the structure and behavior of every part of the Earth, from the earliest fossil-record preserved in rocks to present-day phenomena such as global warming.

The Earth is estimated to be 4,600 million years old. A geologist tries to understand the Earth's past by dating radioactive minerals and rocks, thereby arriving at a chronology of geological events. Many beds of rock contain the remains of shells, bones, or leaves. These objects are called "fossils", a term first used by Georgius Agricola (1494-1555) to mean "dug out of the ground". Since the end of the 18th century, the term has been used only for the relics of animals and plants, which can be traced back to 3,000 million years.

Geology is also useful to industry. The search for fuels has been an important aspect of geological surveys. The search for coal and oil, and later for uranium and other sources of atomic fuel, pushed this interest even further.

## THE SHAPE OF THE EARTH

The recognition of the Earth as a sphere occurred in antiquity and permitted the prediction of eclipses. Following the revival of the idea of heliocentricity (that the sun was the center around which the earth revolved) by Copernicus (1473-1543) in the 16th century, new explorations and scientific reasoning led to a reasonably accurate description of the Earth.

The first voyage around the world was led by Ferdinand Magellan (1480-1521). It began in Seville, Spain, in 1519 and ended in Seville under del Canto (1476-1526) in 1522, establishing beyond dispute that the earth was a globe. Today, it is possible to girdle the earth and to photograph its surface at heights from which the curvature of the globe is clearly visible. Pythagoras of Samos (c. 580-500 BC) was probably the first to consider the possibility that the earth might be a sphere. By observing the approach of ships from beyond the horizon—first the masts and sails and then the hull—Pythagoras realized that the surface of the sea is not flat but curved. Three centuries later, when it was already known that the distance of the sun was so great that the direction of the sun's rays at any moment could be regarded as

## GEOLOGICAL THOUGHT

**c. 350 BC** Theophrastus, Aristotle's successor at the Lyceum, describes several minerals and ores.

**c. 250 BC** Eratosthenes, the chief librarian at Alexandria, estimates the Earth's circumference by studying shadows.

**c. AD 70** Pliny the Elder studies more minerals and identifies the origin of amber as a fossilized resin from trees.

**c. AD 1020** Al Biruni of Persia studies fossils and says that the Indian subcontinent was once a sea.

**AD 1027** Ibn Sina (Avicenna) of Persia formulates the principle of superposition of strata and explains the formation of mountains.

**AD 1696** English theologian William Whiston's *A New Theory of the Earth* argues that the Great Biblical Flood formed the Earth's rock strata.

**AD 1743** French mathematician Alexis Clairault shows the Earth to be an ellipsoid of rotation.

**AD 1749** French naturalist Georges-Louis Leclerc publishes *Histoire Naturelle*.

**AD 1741** The National Museum of Natural History in France designates the first teaching position for geology.

**AD 1774** German scientist Abraham Gottlob Werner proposes that the Earth formed as a precipitate from a great ocean.

**AD 1785** Scottish naturalist James Hutton argues that the Earth is immeasurably old and formed through the gradual solidification of a molten mass.

**AD 1796** French scientist Marquis de Laplace puts forward his 'Nebular' hypothesis, suggesting that the Sun and planets were originally a disc-shaped rotating cloud of hot gas.

**AD 1830** Publication of the first volume of Charles Lyell's *Principles of Geology*, outlining the continuity of geological processes.

parallel, Eratosthenes, the chief librarian at Alexandria, devised a simple and elegant method for estimating the size of the globe. He had heard that at Syene (known today as Aswan), on the banks of the Nile, the sun shone vertically at noon on Midsummer's Day, so that a vertical stick or plumb line threw no shadow. He observed, however, that at Alexandria, roughly 500 miles (800 km) to the north of Syene, there were very perceptible shadows at that time.

The reason for the spherical shape of the earth became clear when Issac Newton (1643–1727) formulated his law of gravitation. All the particles of the earth are pulled towards its center of gravity and a sphere is the natural shape to pack the maximum mass.

The earth is not exactly spherical, however. Again it was Newton who first showed that, because of the earth's daily rotation, its matter is affected not only by inward gravitation, but also by an outward centrifugal force, which reaches its maximum at the equator. He inferred that there should be an equatorial bulge, where the apparent value of gravity was reduced, and a complementary flattening at the poles, where the centrifugal force becomes least. Newton's inference was at variance with the few crude measurements that had then been made. According to these, the earth was shaped not like an orange (with a short polar axis) but like a lemon (with a long polar axis). To settle the matter, the French Academy in 1735 dispatched a surveying expedition to the neighborhood of Chimborazo in the Andes mountains of what is now Ecuador, and followed it

**ISAAC NEWTON (1642-1727)**
Newton's theory of gravitation explained the motion of stars and planets, and the shape of the Earth. Newton's laws of force also proved that the rotation of the Earth on its axis causes an outward centrifugal force that reaches its maximum at the Equator. Therefore, there should be a bulge at the Equator and a complementary flattening at the poles.

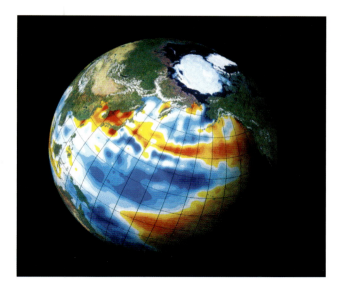

**EFFECTS OF EL NINO IN 1991 AND 1982/83**

Map of sea surface temperature fluctuations in the Pacific Ocean caused by the El Nino phenomenon. In the shaded area at lower right is the warm water of the 1991 El Nino (orange & yellow). The orange and yellow band further north (center) is a sea temperature anomaly caused by the El Nino of 1982-83. The El Nino, along with its atmospheric partner, the Southern Oscillation, results in dramatic climate shifts.

up in 1736 by another to Lapland (now in Finland). The results showed that Newton was right. It is, moreover, significant that before these expeditions returned, the celebrated French mathematician Alexis–Claude de Clairaut (1713-1765) had calculated what the shape of the earth would be, assuming the earth to be a fluid and subject only to the effects of its own rotation and gravitational attraction. The ellipsoid of rotation, now internationally adopted for surveying purposes as most closely representing the shape of the earth, corresponds almost exactly to that calculated by Clairaut.

To sum up, if the surface of the earth were everywhere at sea-level, its shape—the geoid, or the "mathematical figure of the earth"—would closely approximate that of an ellipsoid of rotation (i.e., an oblate spheroid) with a polar axis 7,900 miles

(12,650 km) long, which is nearly 27 miles (43 km) shorter than the equatorial diameter.

How is it, then, that the earth is not exactly a spheroid? The reason is that the rocks that form the earth's crust are not of the same density throughout. Since the equatorial bulge is a consequence of the relatively low value of gravity around the equatorial zone, it follows that there should be bulges in other places where gravity is relatively low—that is to say, wherever light, sialic rocks make up an appreciable part of the crust. Such places are the continents. On the other hand, wherever the crust is composed of heavy rocks, gravity is relatively high and the surface should be correspondingly depressed. Such regions are the ocean basins.

## THE CHANGING EARTH

The recognition of radioactivity in rocks enabled geologists to determine that the Earth is billions of years old. New measurements from the ocean bottom and from ice cores supported the theory that ice ages are caused by periodic and predictable changes in the Earth's orbit. Weather satellites began to produce global data that led to significant improvements in weather forecasts. An area of intense interest over the past two decades has been the interactions of the oceans and the atmosphere that produce major climate anomalies or planetary oscillations, such as El Nino and its atmospheric partner, the Southern Oscillation.

The past few decades have also seen a revolution in our appreciation of the chemistry of the global environment and its susceptibility to change. This revolution has been generated by the recognition that the Earth's protective ozone layer is controlled by the complex interplay of atmospheric chemistry and circulation, and that increases in trace gases (e.g., the chlorofluoromethanes, or CFMs) could seriously deplete the ozone layer over the next

**THE BLUE PLANET**
Satellite image of the Earth showing North and South America and the Pacific Ocean. The image was obtained by one of the Geostationary Operational Environmental Satellites (GOES) of the US National Oceanic and Atmospheric Administration (NOAA))

**OZONE HOLE RESEARCH, ANTARCTICA**
Scientists on the Ross Ice Shelf launching a helium balloon to carry measuring instruments into the atmosphere above Antarctica. The instruments will measure temperature, pressure, and position near the edge of the southern polar vortex to assess the erosion of the ozone layer. A depleted ozone layer would cause temperatures on the Earth to rise at faster rates. Photographed in October 2005.

decades. More recent research has shown that concentrations of the important "greenhouse gases"—methane and nitrous oxide—are increasing at alarming rates.

## EARTH SYSTEM PROCESSES

Two primary conclusions have emerged from contemporary research in the Earth sciences. First, change on a planetary scale is the result of interactions among the Earth's different subsystems—the atmosphere, ocean, mantle and crust, cryosphere, and biological systems. Moreover, change on any temporal scale involves interactions among Earth system processes that occur over diverse time scales. The traditional Earth sciences have, in general, been concerned with structure and process within specific subsystems and within specific temporal ranges. To study change on a planetary scale, we need to integrate the efforts of the Earth sciences and take a broader, global view. This is the task of Earth system sciences.

Consider Earth from the distant view made possible by space technology. The planet we see—clouds swirling over blue

oceans—has been properly referred to as the "Blue Planet". How might we describe the processes and the evolution of this object? Ideally, we seek to represent its state at any time by a collection of variables and to determine how each of these variables will change with time—that is, we seek to treat the Earth as a dynamical system. Clearly, a dynamical system that would represent all processes and change on the Earth would be an immensely complicated mathematical structure, which is at present beyond our powers to develop. In practice, then, we often choose to modify that dynamic order to examine processes on specific time scales.

Five distinct bands can be used to define the major time scales involved in global change. These bands are:

Millions to Billions of Years: The Earth formed rapidly, within a hundred million years, and since that time a metallic core (responsible for the magnetic field) has remained largely isolated from the overlying mantle and lithosphere. The characteristic time scale for the operations of these systems is millions of years; the evolution of life and the associated development of the present chemical composition of the atmosphere occurred over similar time scales.

Thousands of Years: The oscillations between ice ages and interglacial periods (with associated variations in atmospheric chemical composition), the development of soils, and the distribution of biological species occurred largely in response to changes in the Earth's orbit around the sun that recur in cycles of tens of thousands of years.

Decades to Centuries: The changes that threaten the viability of some forms of life on the planet—changes in climate, chemical composition of the atmosphere, patterns of surface aridity or acidity, and in terrestrial and marine biological systems—occur over decades and centuries.

Days to Seasons: Weather phenomena, eddies in ocean currents, seasonal growth and melting of the polar sea-ice covers, surface runoff and weathering, and the annual cycle of plant growth are all confined to time scales regulated by the annual cycle of insolation (the measure of solar radiation energy). A large part of the feedback from the biogeochemical cycles occurs through the alteration of radiative processes that supply energy to the major subsystems. Earthquakes and volcanic eruptions are episodic manifestations of adjustments taking place within the solid Earth over much longer time scales; the sudden violence of such cataclysmic events obscures the fact that tens to hundreds of years are required for accumulation of the energy needed for a repetition of the event.

Seconds to Hours: Changes occurring among land, oceans, the atmosphere, and the biota (living forms) are all dominated by processes with time scales shorter than a day.

Geology is a relatively recent subject. In addition to its core branches, advances in geology and allied fields have led to specialized sciences like geophysics, geochemistry, geohydrology, glaciology, seismology, oceanography, rock mechanics, photogeology, and remote sensing. The science of geology has been further subdivided into the following branches for the sake of systematic study: physical geology, mineralogy, petrology, structural geology, historical geology (stratigraphy), paleontology, and economic geology.

For geologists, a rock is more than an assemblage of minerals: in it is inscribed the history of the earth. Although geology has its own laboratory methods for studying mineral, rock, fuels, etc., it is essentially a field science. It leads geologists to waterfalls, glaciers, volcanoes, beaches, coral reefs, and even down to the sea floor in search of geological events and data. What can be more exciting than this first-hand observation!

**CAVE OF CRYSTALS**
Geologists in the Cave of Crystals (Cueva de los Cristales) in Naica Mine, Chihuahua, Mexico. The crystals are the largest known in the world and formed over millions of years in the mineral-rich, geothermally heated water that filled the caves. The Cave of Crystals is 951 feet (290 metres) deep, and was discovered in 2000. Geology is essentially a field science and recent discoveries such as these prove that much of the world still lies unexplored.

# THE BUILDING BLOCKS

In ancient times, astronomers noticed how certain lights moved in the sky in relation to other fixed celestial bodies. The ancient Greeks called these lights *plantetoi* (wanderers). Astrologers held that the movement of these planets affected the lives of humans. The belief held true for almost two millennia, until astronomers found that the Earth and the Sun were just one of billions of stars and planets in the Universe.

## THE BIRTH OF STARS AND PLANETS

Most theories for the origin of the planets assume that the planetary material was derived from the Sun, or from a primitive cloud of dust and gas that later condensed. In general, the theories can be divided into two groups:

(a) Natural or evolutionary theories, which suggest that planetary systems form as part of the evolutionary history of stars. If these theories are correct, planetary systems around stars should be relatively numerous.

(b) Catastrophic theories, which imagine that the planetary systems form only by some special accident or catastrophe, such as the close approach or collision of two stars. However, because the stars are so far apart in the galaxy, the possibility of such a catastrophe is extremely remote. The probability is so low that only about one in ten thousand million stars can acquire a planetary system by this mechanism during the entire life of the solar system ($4.6 \times 10^9$ years).

Among the many theories for the origin of the solar system, perhaps the best known is that put forward by the Marquis de Laplace (1749–1827) in 1796. The Laplace or 'Nebular' hypothesis suggested that the material that formed the Sun and planets was originally a disc-shaped rotating nebula, or cloud of hot gas. As the gas lost energy by radiation and became cooler, the cloud would have shrunk inwards, and would have rotated more rapidly to conserve angular momentum, until, at some period in the contraction, the speed of the outermost rim of the disc would have become sufficient for the 'centrifugal force' to be as great as the inward gravitational attraction. At this position, the material would be in stable orbit so that continued contraction of the cloud would have left behind a continuous ring or disc of material. The vast mass of the nebula would have continued to shrink and finally formed the Sun. In this way, one could explain the regularity of movement within the bodies of the solar system.

**NEBULAR FORMATION.** Artwork representing the formation of a planetary system around a star based upon the nebular hypothesis.

After an initial period of wide acceptance, this hypothesis fell into disfavor because it could not explain the energy distribution within the solar system. The planets have only 0.1 per cent of the solar system's mass, but 98 per cent of the energy of movement, i.e., the energy of angular momentum. Until recently, there seemed no obvious way for this energy to be lost. Because of this difficulty, the catastrophic theories became popular. These explained the very high energy of movement of the planets as being imparted not from the Sun but by the close approach of another star. If the star had come very close to the Sun, its tidal attraction on the Sun would have caused some of the Sun's mass to be torn away from the surface. This ejected material, under the influence of the gravitational attraction of the other star, instead of falling back into the Sun, assumed a path around it and, on condensation, formed the planets. Because the passing star imparted the direction of movement, all of the ejected material would have moved in the same direction around the Sun.

Since about 1943, there has been a swing back to theories of a Laplacian type, in which the planets formed from a rotating gaseous nebula or dust cloud, either around the Sun and perhaps derived from it, or else parental to the Sun, which formed during the condensation processes. The objection to Laplacian theories that the Sun's angular momentum is too small has now been overcome. New calculations take into account a host of factors such as the sun's magnetic dipole moment, the different spin axes of planets and their satellites, the presence of dark matter, etc., to explain the conservation of angular momentum in the solar system.

Specific details of the process of planetary formation from the initial cloud of gas and dust, are as yet uncertain. Studies of the chemical nature of planetary accretion show that the inner planets formed from solid materials at relatively low temperatures. The relative paucity of the inert gases—helium, neon, argon, xenon, and knypton—in the Earth compared with the Sun indicates that gases did not accrete to form the earth. The present oceans and atmosphere of the Earth are secondary features resulting from the subsequent de-watering or 'sweating-out' of the Earth's interior. The original water would have accreted in the Earth not as gas but in solid hydrated minerals such as amphibole or mica.

### AVICENNA (981-1037)

Avicenna's Arabic name was Abu Ali Al Husain Ibn Abdullah Ibn Sina. He was a philosopher and scientist of amazing versatility, writing on astronomy, alchemy, medicine, and minerals, among other things. He was one of the earliest to theorize about the formation of mountains and seas, and about weather phenomena.

**HISTOIRE NATURELLE** Animals being dissected and examined under a microscope. Illustration from Histoire Naturelle by Georges Louis Leclerc, Comte de Buffon (1707-1788). Buffon formulated one of the earliest theories of evolution.

**LEFT:** The Deluge, Genesis 7:20-24. Illustration from Dore's *The Holy Bible*, 1866 (engraving).

## ORIGIN OF THE EARTH

Medieval Arab scholars had been grappling with the processes that had formed the Earth the way it was. Al Biruni (973–1048) was one of the earliest Muslim geologists, whose works included the earliest writings on the geology of India, hypothesizing that the Indian subcontinent was once a sea. Ibn Sina or Avicenna (981–1037), a Persian polymath, wrote an encyclopedic work in which he theorized on the formation of mountains, the advantages of mountains in the formation of clouds, sources of water, origin of earthquakes, formation of minerals, and the diversity of the Earth's terrain.

But in Europe, due to the strength of Christian beliefs during the 17th century, the theory of the origin of the Earth that was most widely accepted was *A New Theory of the Earth*, published in 1696 by William Whiston (1667–1752). Whiston used Christian reasoning to "prove" that the Great Flood had occurred and that the flood had formed the rock strata of the Earth.

In 1749, the French naturalist Georges-Louis Leclerc, Comte de Buffon (1707–1788) published his *Histoire Naturelle* in which he attacked the popular Christian concepts of Whiston and other Christian theorists on the topic of the history of Earth. From experimentation with cooling globes, he found that the age of the Earth was not 6,000 years as stated in the Bible, but rather 75,000 years.

Abraham Gottlob Werner (1749–1817), a German scholar, reached a very successful compromise when he theorized that all rocks were the direct result of either of two processes: (1) deposition in the primeval ocean, represented by the Noachian flood, or (2) sculpturing and deposition during the retreat of this ocean. Werner's interpretation came to represent the 'Neptunist' conception of the Earth's beginnings.

floor provide satisfactory values for the increase of temperature with depth near the Earth's surface, and for the 'heat flow'—the rate of escape of interior heat from the Earth's surface. The temperature of the Earth's interior is defined to some extent by the requirement that the oxides of the lower mantle are still not molten, and the core changes from solid to liquid, i.e., melts, at about 3,220 miles (5,150 km) or at over 3 million atmospheric pressure. By comparing laboratory experiments on the effect of pressure on the melting point of similar materials, it seems that the Earth's core is likely to be at temperatures around 4000 K (3727°C). Other sources of evidence support temperatures of this order in the core.

In the late 1780s, the Scottish scientist James Hutton (1726–1797) launched an attack on the Neptunist approach, arguing that the Earth could not be viewed as static and bereft of continuous change. Hutton formulated the principle of uniformitarianism, which holds that Earth processes occurring today had their counterparts in the ancient past. The publication of Hutton's two-volume *Theory of the Earth* in 1795 firmly established him as one of the founders of modern geologic thought. Werner's 'Universal' and 'Partial' categories of rock layers were discredited when Hutton observed that basaltic rocks on the outskirts of Edinburgh could not have formed as a precipitate from the primordial ocean, as Werner had claimed. Furthermore, the basalt probably came from the Earth's interior, a process in clear conflict with Neptunist theory. Hutton found a brilliant disciple in his fellow countryman, Charles Lyell (1797–1875), who established uniformitarianism as a fundamental principle of geology.

## THE TEMPERATURE OF THE EARTH

A good deal is known about the present temperatures within the Earth. Temperature and conductivity measurements in deep boreholes and mines and from sediments in the ocean

From a geological point of view it is important to know how long these temperatures have existed. Did the Earth begin as a cool body and increase in heat, or has it always had temperatures

**FORMATION OF THE SOLAR SYSTEM**

Artwork showing the different sizes of objects (right-hand column) in the solar system at different stages (left-hand column) during its formation. Some 5 billion years ago (top), the primordial solar nebula, a cloud of gas and dust particles, began to gravitationally accrete to form larger particles that orbited the newly formed Sun. The Earth formed 4.5 billion years ago. The composition of the particles varied, from icy particles (top) to rocky planets with metallic cores (bottom).

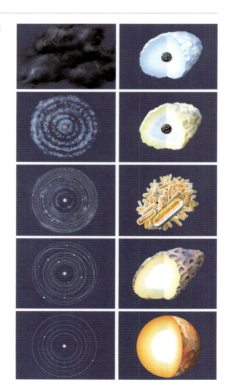

of this order? This is particularly important because the convective movement of the mantle, which causes seafloor spreading and continental drift, requires heat for its operation, and would increase in speed if the temperature increased and the viscosity of the Earth's interior diminished. Also, does the lack of igneous (or any other) rocks older than 4,000 million years and the gap of 600 million years between that time and the Origin of the Earth (4.6 x 10$^9$ years, or 4,600 million years, ago) result from the fact that early temperatures were so low that igneous processes did not occur? Or the early temperatures were so high that rapid convective movement engulfed and destroyed the earliest surface rocks?

If, as suggested earlier, the Earth accumulated from solid particles at temperatures of about 500 K or less, how do we explain the presence of molten rocks in the volcanoes? This problem was one of the early reasons for supposing that the planets were derived from the Sun. The Sun was so hot that one could easily imagine the Earth's present heat to be a residue of this original solar temperature. We know now that the Earth has its own heat source: the radioactive decay of elements such as uranium and thorium and their daughter products. These are a considerable heat source. For example, in one year a kilogram of uranium and the equivalent amounts of its decay products would emit 3.1 x 10$^3$ joules of heat. Moreover, such elements as potassium and rubidium, not usually thought of as radioactive, are in fact very weakly radioactive because they contain small amounts of the isotopes K40 and Rb87 and although the energy output of potassium is very low (1 kg of potassium emits about 0.1 joules per year) the abundance of potassium in the Earth is so much greater than that of uranium that potassium is almost as important a source of radiogenic energy. Although there is no certainty about the total concentrations of radioactive elements within the Earth, there is no question that these can provide the necessary heat to explain the Earth's present temperatures.

There are other sources of heat. As the Earth grew and its gravitational attraction became greater, the energy released by an in-falling body increased. At present, the energy of infall of a meteorite or a space rocket is very large, enough to cause complete vaporization under some circumstances. For the whole Earth, the gravitational energy release during accumulation and formation is equivalent to 38 x 10$^6$ joules per kilogram, enough to raise the temperature of the Earth by more than 20000°C.

Another source of energy is also gravitational. If the Earth accumulated as a homogeneous mix of metal and silicate, and subsequently the metal phase sank down to the center to form the core, this redistribution of lighter and denser phases within the Earth would turn a considerable amount of potential energy into heat, again enough to cause the core to melt.

**ERUPTION OF VESUVIUS IN AD 79**

The Roman city of Herculaneum is destroyed by the eruption of the Mount Vesuvius volcano. Pliny the Elder, who was in command of a Roman fleet, died along with several of his men when he moved his ships closer to observe the eruption. More than 500 million people live in regions vulnerable to volcanic destruction.

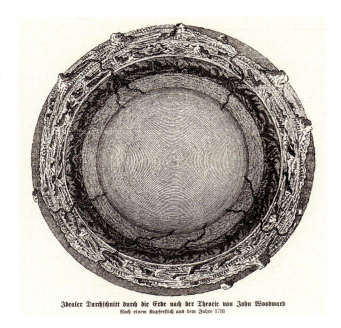

Idealer Durchschnitt durch die Erde nach der Theorie von John Woodward
Nach einem Kupferstich aus dem Jahre 1735

# THE CRUST

The Earth is estimated to be 4.6 billion years old. Its radius at the equator is approximately 4.000 miles (6,400 km) and total surface area is 197 square miles (510 million sq km), of which dry land accounts for 58 square miles (150 million sq km), or 29 per cent.

All the planets of the solar system seem to have had similar histories: they condensed from a nebula; each has scars caused by a rain of meteorites; and most have had or have a thermal history that produced magma, molten rock, and dissolved gases (familiar on Earth as lava), resulting in volcanoes and mountains. But there are vital differences between the planets: for example, the Earth's atmosphere and hydrosphere are unique.

Organic remains provide evidence that primitive life existed on the Earth at least 3.5 billion years ago. Living organisms have been an essential factor in the development of the Earth's

**EARTH'S INTERIOR**

English geologist John Woodward's (1665-1728) conception of the composition of the globe, with molten lava surrounded by a thick crust (1725). Woodward showed that the stony surface of the earth was divided into strata, and that the enclosed shells were originally generated in the sea; but his explanations for the formation of rocks were erroneous.

crust over the greater part of its history, clearly differentiating the Earth from the other planets.

The thickness of the Earth's crust varies between 12 miles (20 km) and 44 miles (70 km) on the continent and between 3 miles (5 km) and 9.5 miles (15 km) under the oceans. The density of the crustal rocks averages between 2.0 and 2.7 g/cm$^2$ and the velocity of longitudinal elastic waves is usually 3.7–4.5 miles per second (6–7.4 km/sec). The crust can be subdivided into what are known as sedimentary, granitic, and basaltic layers, according to their seismic characteristics. The crust is separated from the mantle by the Mohorovicic (or 'Moho') discontinuity, a seismic boundary at which the velocity of elastic waves increases markedly to values of more than 5 miles/second (8 km/sec), corresponding to an increase in the density of the material to 3.3-3.7 g/cm$^2$. The mantle can be subdivided into two unequal parts, an upper mantle to a depth of 250 miles (400 km) and a lower mantle to a depth of about 1,810 miles (2,900 km). Geophysical evidence has identified a layer within the upper mantle, which is more plastic, less viscous, and therefore more mobile than the layers above and below. This layer is called the asthenosphere.

The lithosphere is the "stony cover" of the Earth. Together, the lithosphere and the asthenosphere are often called the tectonosphere, emphasizing their leading role in the genesis of tectonic movements and structures on the Earth's surface. The structural and material heterogeneity of the lithosphere is not reflected in the remainder of the mantle and the core, which are characterized by a high degree of horizontal homogeneity and by a more gradual change of physical properties with depth.

The lithosphere is thin compared with the other layers and has a variable thickness. Below the continents, it reaches depths of up to 50 miles (80 km), and according to some data even up to 75 miles (120 km), while under the oceans its thickness amounts to only 1.9 miles (3 km) or less.

Minerals, usually in combination, form the three classes of rocks: igneous (solidified from a molten state), sedimentary (formed by the erosion of existing rocks followed by re-deposition), and metamorphic (formed by the action of heat and pressure on existing rocks). Almost 2,000 minerals are known and they show a great diversity of composition. Minerals that form rocks are a small group. Of these, the silicates are the most important, followed by the two carbonates, calcite and dolomite. Such minerals as oxides and sulphides, many of which are economically valuable as ores, do not always form large masses of rock. Only about 50 minerals are abundant enough to make up any substantial part of the Earth's crust.

**CONTINENTAL DRIFT** Model, globe of the Earth during the Cretaceous period, time of the dinosaurs with the continents scattered in unfamiliar shapes and locations.

The southern part of Africa floats alone in huge prehistoric sea. While the theory of continental drift was first proposed in 1912, it was only in the 1960s that the theory of plate tectonics came up with an explanation of continental motion through the mechanism of seafloor spreading. New rock is created by volcanism at mid-ocean ridges and returned to the Earth's mantle at ocean trenches.

Most minerals are produced from, or by, cooling magma or lava; subsurface solutions involving hot water or hot gases, including steam; hot volcanic vapors; chemical reaction with earlier minerals; replacement of an earlier mineral; re-crystallization; or evaporation of solutions in water.

## CONTINENTAL DRIFT

Since the end of the 1960s, the geological sciences have been guided by a theory which explains the movements of the lithosphere and the deformation of the Earth's crust that have produced fundamental changes in the relief of the Earth's surface, in the distribution of dry land and water, mountains and plains. This important concept is the theory of plate tectonics. It is derived from the theory of Continental Drift proposed by the German scientist Alfred Wegener (1880–1930) in 1912. Wegener's theory suggested that the continents were joined together at a certain time in the past and formed a single landmass known as Pangaea; thereafter they drifted like rafts over the ocean floor, finally reaching their present position. The shapes of continents and matching coastline geology between some continents attested to this splitting.

**ALFRED WEGENER (1880-1930)**
Impressed by the matching coastlines of South America and Africa, German geophysicist Alfred Wegener suggested in 1912 that there had once been a single supercontinent (which he called Pangaea) that had fragmented and drifted apart into the continents we know today.

In addition to the vertical layering of the Earth, there are lateral changes in the crust. These are connected with the history of the formation of the Earth's layers over billions of years.

The plate tectonics theory was developed by earth scientists who brought together data from terrestrial magnetism, seismology, and of course geology, to explain the formation of the oceans and the fold-mountain chains, as well as volcanic activity, earthquakes and a number of other dynamic features. According to it, the lithosphere, the crust, and the adjacent part of the mantle, are divided into a small number of large 'plates' that move substantial horizontal distances (and smaller vertical distances) with respect to each other. The displacements of the plates, in response to movements in the underlying mantle caused by convection currents, lead to changes in the relief of the Earth and the distribution of the continents.

The continents and oceans are distributed unevenly: about 70 per cent of the land is concentrated in the northern hemisphere, and 75 per cent of the land lies between the longitudes 150°

**VULCAN**

Volcanoes derive their name from Vulcan, the Roman god of fire. This tapestry painting from 1757 (oil on canvas) depicts Vulcan presenting arms for the Trojan hero Aeneas to his mother Venus. Vulcan is also identified with Hephaestes, the Greek god of smiths.

West and 105° East. This uneven distribution is known as the equatorial dis-symmetry. All the continents except Antarctica have a wedge-shaped form, narrowing to the south. Accordingly, the northern limits of the oceans also have wedge-shaped outlines. The remarkable dovetailing of the coastal outlines of several continents is familiar, particularly of the eastern coast of South America and the western coast of Africa. The geological structures of the coastal areas also correspond.

## GEOLOGICAL 'DISTURBANCES'

There are two major groups of geological processes. One consists of internal (endogenous) processes, such as volcanic eruptions, faulting and earthquakes, deformation of large rock masses, metamorphism, and the de-gasification of the

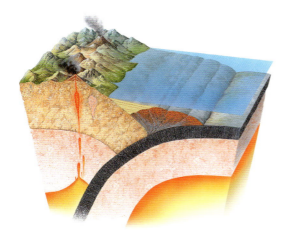

**ACCRETIONARY PRISM**

A cross-section through the Earth's crust showing the formation of an accretionary prism at a subduction zone. A subduction zone occurs where two convergent tectonic plates collide.

Earth's interior. The other group includes external (exogenous) processes, which range from weathering, abrasion, erosion, and deposition of sediments, to landslides, rockfalls, and mudflows. Both the internal and external processes determine the natural environment and many represent natural hazards that can cause considerable damage.

Of the deep-seated processes that influence the natural environment, volcanoes and earthquakes are the most dangerous, particularly as more than 500 million people live in regions vulnerable to them. Volcanoes take their name from Vulcan, the Roman god of fire. A volcano is the vent from which molten magma (lava on the surface), solid rock debris, and gases are erupted. The erupted lava, ashes, and other rock debris construct new landforms adjacent to the vent.

Volcanic eruptions can be classified broadly as non-explosive or explosive. Non-explosive eruptions are generally of an iron and magnesium rich magma that is relatively fluid and allows gas to escape readily. Eruptions of silica-rich magma are usually violent, as it is not so fluid. The gases are released explosively, often accompanied by vast clouds of ashes (tuft).

Earthquakes, sudden movements of the ground, are caused by an abrupt release of energy that has accumulated in the underlying rocks. There are millions of earthquakes throughout the world every year, varying from minor tremors recorded only by sensitive instruments (seismometers), to a few great earthquakes that cause enormous damage and loss of life. Although there is no agreement among scientists about the precise mechanism that triggers an earthquake, it is generally accepted that the release of the stored shock energy is brought about by a sudden disturbance of the continuity of the previously stressed rocks. There is clear evidence that these stresses are built up over a long time and that they can affect very large regions.

Among the external processes, landslides, and mudflows can have catastrophic effects. Both can be on a particularly large scale if strong earthquakes occur when the rocks are wet. This applies particularly to clay deposits, most of which lose strength if they are shaken when wet. Mudflows are sudden and short-lived, usually during heavy rains or intensive melting of snow and ice (as may be caused by a volcanic eruption) or caused by the breaching of dams in valleys where there are large supplies of loosely fragmented material.

Weathering is the irreversible change in the composition, structure, and properties of rocks as a result of the action of climatic and biological factors. Under their influence, hard rocks

**EARTHQUAKE IN SAN FRANCISCO, 1906**

A synagogue in San Francisco after the 1906 earthquake. Earthquakes, sudden motions of the ground, are caused by an abrupt release of energy that has accumulated in the underlying rocks. The release of the stored shock energy is brought about by a sudden disturbance in the continuity of rocks that are under stress.

D. Nummulitenkalkplatte.

1. Nummulites nummularia von oben. 2. Horizontal durchschi

Nummulites planulata
von oben.          von der Seite.

5. Robulina echini

6.7. Rotalia Partschiana.          8. Amphistegina Hai

E. Nummuliten und Verwandte.

B. Delesserites Gozolanus.

C. Kieselschieferplatte mit Abdrücken.

1. 2. Blätter und Frucht von Podogonium Knorrii.   3. Blätter von Cinnamo-
mum Scheuchzeri.   4. Blatt von Andromeda protogaea.   5. Blatt von Sapindus

are transformed into rock debris and clay. The effect is greatest close to the surface, but rocks normally can be weathered to a depth of 82 feet (25 meters) and to more than 328 feet (100 meters) if they have been deeply fractured by tectonic processes. Weathered rocks, because of their weakened state, are particularly susceptible to landslides.

Erosion is the destruction of rocks and the transport of the material produced, commonly by running water, but also by wind and ice. It is a widespread process: for example it has been estimated that the area under erosion in the former USSR amounts to 6.5 million hectares. In India, the area subject to severe erosion is 150 million hectares, almost half the territory for which information exists.

In cold climates the freezing and thawing of ice in crevices has a very destructive effect on rocks. The presence of permafrost (permanent ice below the ground) can also have a destructive consequence and adversely affect the construction of buildings, roads and other engineering structures, and the laying of pipelines. The engineering structures themselves can activate such cryogenic processes.

## ROCKS THAT DEPICT THE PAST

Rocks are the pages of the Earth's history book. The fundamental principle involved in reading their meanings was first enunciated by James Hutton in 1785, when he declared that ' the present is the key to the past'—meaning that 'the past history of our globe must be explained by what can be seen to be happening now.' Rocks and characteristic associations of rocks, with easily recognizable peculiarities of composition and structure, are observed to result from processes occurring at present in particular kinds of geographical and climatic environments. If similar rocks belonging to a former geological age are found to have the same peculiarities and associations, it is inferred that they were formed by the operation of similar processes in similar environments.

The presence of fossil corals or of the shells of other marine organisms in a limestone indicates that it was deposited on the sea floor, and that what now is land once lay beneath the waves. The limestone may pass downwards or laterally into shale, sandstone, and conglomerate. The last of these represents an old beach, and indicates the shore line.

Elsewhere, old lava flows represent the eruptions of ancient volcanoes, and vents that were active many millions of years ago still figure prominently in the landscape at some places. Beds of rock-salt point to the former existence of inland seas that evaporated in the sunshine. Seams of coal, which are the compressed remains of accumulations of peat, suggest widespread swamps and luxuriant vegetation. Smoothed and striated rock surfaces associated with beds of boulder clay prove the former passage of glaciers or ice sheets. In every case, the characters of older formations are matched with those of rocks now in the making.

The stratigraphical column can be defined as the sequence of rock formations arranged according to their order of formation in time. William Smith (1769–1839), Georges Cuvier (1769–1832), and Alexander Broignart (1770–1847) can all be recognized for their roles during the early 19th century in furthering the concept of fossil-based stratigraphy. Another important name is that of Adam Sedgwick (1785–1873), who in 1833 mapped rocks that he had established to be from the Cambrian Period. The Cambrian is the earliest period in whose rocks are found numerous large, distinctly fossilizable multicellular organisms. This sudden appearance of hard body fossils is referred to as the Cambrian explosion.

The stratified rocks accumulated layer upon layer, and where a continuous succession of flat-lying beds can be seen, such as on the slopes of the Ingleborough mountain in Yorkshire, where any inversion of beds by overfolding or repetition of beds by overthrusting never occurred, it is obvious that the lowest beds are the oldest and those at the top of the series the youngest. The Grand Canyon in the American state of Colorado presents one of the finest successions of this kind.

## THE PRINCIPLE OF SUPERPOSITION

In 1669, the Danish-born natural scientist Nicolaus Steno (1638-1686) published his dissertation *Prodromus*. This work laid down the essential method of geology by showing that layered rocks exhibit sequential change—that they contain a record of past events. Steno concluded from his observations that rocks deposited first lie at the bottom of a sequence, while

**WILLIAM SMITH (1769-1839)**

The son of a blacksmith, Smith became a canal surveyor, which allowed him to study the varied geology of England and Wales. He found that geological strata could be reliably identified at different places on the basis of the fossils they contained. In 1815, he published the first geological map of England and Wales. He is considered the father of English geology.

**FAR RIGHT: ADAM SEDGWICK (1785-1873)**

In 1835, Sedwick worked out the stratigraphic succession of fossil-bearing rocks in Wales, naming the oldest of them the Cambrian period (now dated at 500-570 million years ago).

**JOHANN GOTTLOB LEHMANN (1719–1767)**

Lehmann, a German geologist, studied the succession of rocks in the southern part of his country and in the Alps, and classified them into three distinct layers. His scheme of classification was widely used throughout Europe and Lehmann became one of the founders of the science of stratigraphy.

*D. Johannes Gottlob Lehmann*

those deposited later are at the top. This is now known as the principle of superposition. In the 11th century, Avicenna of Persia had formulated a similar theory of superposition.

Steno's ideas spurred naturalists around Europe. In England in 1725, the early geologist John Strachey (1671–1743) produced probably the first modern geologic maps of rock strata. In 1756, Johann Gottlob Lehmann (1719–1767) of Germany reported on the succession of rocks in the southern part of his country and the Alps, and classified them into three distinct layers. Lehmann's threefold classification scheme was successfully applied with minor alterations to studies in other areas of Europe. By the latter part of the 18th century, the superpositional concept of rock strata had been firmly established through a number of independent investigations throughout Europe.

**CHARLES LYELL (1797-1875)**

Lyell's three-volume masterpiece Principles of Geology (1830-33) conclusively proved James Hutton's principle of Uniformitarianism. He was also a great influence on Charles Darwin.

## THE SIGNIFICANCE OF FOSSILS

Fossils may be defined as the relics and remnants of ancient animals and plants that have been preserved inside rocks by natural processes. Occurrence of fossils in sedimentary rocks is a matter of chance because they occur only under very favorable conditions. This means all sedimentary rocks do not possess fossils. The fossils and the rocks that possess them belong to the same age, i.e., the rocks that had formed in a particular geological period will have the relics of only those animals and plants which existed at that time.

Around the time Charles Lyell was trying to understand the past through the present, a study of the fossiliferous strata of the Paris Basin brought evidence that rock successions were not necessarily complete records of past geologic events. In fact, significant breaks frequently occured in the superpositional record that also affected the character of the fossils found in the various strata.

**CHARLES DARWIN (1809–1882)**

British naturalist and founder of the theory of evolution, Charles Darwin.

**LEFT:** Georges Cuvier (1769–1832), the French naturalist, lecturing on paleontology at the Museum of Natural History in Paris.

An 1812 study of the Paris Basin by the French zoologist Georges Cuvier showed that many fossils, particularly those of terrestrial vertebrates, had no living counterparts. Indeed, they seemed to represent extinct forms, which, when viewed in the context of the succession of strata, constituted a record of biological succession punctuated by numerous extinctions. Such extinctions, in turn, were followed by a seeming renewal of more advanced but related forms and these were separated from each other by breaks in the associated rock record. Many of these breaks suggested "catastrophic" events. Cuvier explained this by invoking a sequence of recurring catastrophic geologic events, which contributed to massive extinction of species, followed by biological renewal.

The French biologist Jean-Baptiste de Monet, Chevalier de la Marck (Lamarck) (1744–1829) disagreed with Cuvier's interpretation and preferred to think of organisms and their distribution in time and space as responding to the distribution of favorable habitats. If confronted with the need to adapt to abrupt changes in local habitat—Cuvier's catastrophes—species must be able to change in order to survive. If not, they became extinct. Lamarck's approach, much like that of Hutton, stressed the continuity of processes and the continuum of the stratigraphic (rock layer) record.

Evolutionary theories were turned on their head when Charles Darwin (1809–1882) explained the extinction of species by his theory of evolution by natural selection, published in *On the Origin of Species* in 1859. On his voyage aboard the HMS *Beagle*, Darwin came across a bewildering diversity of fossils and living forms that showed both continuity and variation. Darwin inferred that species keep evolving into new ones to survive changes in habitat, climate, etc., and those that fail to keep pace with evolutionary competition become extinct.

# Chapter 3
# WHAT LIES BENEATH

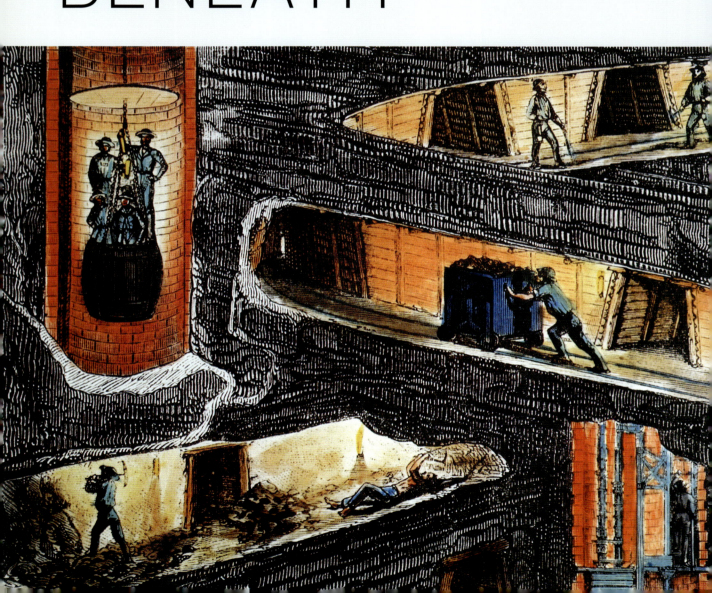

**THE MINE, 1572**

An oil on slate painting showing workers in a coal mine in Europe in the 16th century.

**FACING PAGE** Artwork depicting the cross-section of a coal mine in Europe in the early 19th century, the period when mining became one of the biggest industries in Europe.

The 20th century saw two great changes in the field of geology. As the industrial economies expanded and new technologies developed, geology was increasingly harnessed to the search for resources hidden in the earth. Intensive explorations for minerals, hydrocarbons, metals, gases, and radioactive material were undertaken both at land and sea. So great and rapid has been the mining of minerals that more of them have been extracted in the past 50 years than in all preceding history. With the increasing growth of the construction industry, the need for building material has meant that billions of tons of minerals used in the manufacture of cement and steel are extracted every year. If archeologists have divided the past periods of human history into the Stone Age, the Bronze Age, and the Copper Age, it wouldn't be inappropriate to term the modern period as the 'Cement Age'.

The second great change marked a paradigm shift in the understanding of geological processes. How are landforms created, what causes earthquakes and volcanoes, are land and sea changing with time? The answers to these and several other puzzling questions came with a theory whose role in the science of geology has been compared to the Copernican revolution in astronomy. This theory was of plate tectonics, and,

as noted in the previous chapter, developed from the ideas of continental drift proposed by Alfred Wegener and from the later theory of seafloor spreading, which extended Wegener's ideas and was empirically established in the mid 20th century. The propositions of plate tectonics have also helped shape the science of deep sea marine geology and, consequently, the industry of deep sea ocean mining since the 1960s.

## THE IDEAS OF ALFRED WEGENER

It was known for centuries that the west African coastline and the east coast of South America and the Caribbean Sea fitted together like two pieces of a jigsaw puzzle. A similar fit also appeared across the Pacific. The fit is even more striking when the submerged continental shelves are compared rather than the coastlines, though this discovery was made only in the 20th century. Searching for evidence to further develop his theory of continental drift, which suggested that all continents were part of one landmass in the past (which he called the Pangaea, or 'All-Lands'), Wegener came across an academic paper suggesting that a land bridge had once connected Africa with Brazil. This proposed land bridge was an attempt to explain the well-known observation that the same fossilized

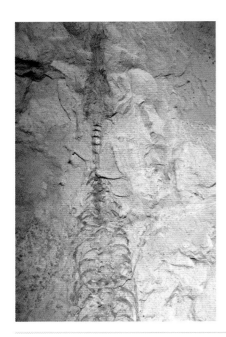

**EVIDENCE OF CONTINENTAL DRIFT**

A fossil of the freshwater reptile *Mesosaurus brasilensis*, found only in western parts of Africa and eastern parts of South America. Such fossils provided evidence for Alfred Wegener's hypothesis that the continents had once been joined together.

The same scraping patterns can be found along the coasts of South America and South Africa. The South African geologist Alexander du Toit (1878–1948) was one of Wegener's early supporters and played a considerable role in the popularization of his ideas. But Wegener's theory was never formally accepted during his lifetime.

## FORMATION OF MOUNTAINS

Wegener's drift hypothesis also provided an alternate explanation for the formation of mountains, a field of study called orogenesis. In the 19th century, the phenomenon of mountain formation was explained through the Geosyncline theory, which was also called the "Contraction theory". It suggested that the earth was once a molten ball and was now cooling. This cooling led to contraction and caused the surface to crack and fold on itself, pushing up the land above the cracks. The problem with this theory was that all mountain ranges should then be of approximately the same age, and this was known to be not true.

plants and animals, belonging to the same time period, were found in South America and Africa. This was also for fossils found in Europe and North America, and in Madagascar and India. Many of these organisms could not have traveled across the vast oceans that currently exist. Wegener's drift theory seemed more plausible than the idea that land bridges once connected all the continents.

But Wegener needed more proof to support his idea. This came through the discovery of identical glacial deposits in different continents. Fossils found in glacial deposits in India, for example, belonged to the same period of 300 million years ago as the remains found in glaciers in the Arctic. Marks left by the scraping of glaciers over the land surface indicated that Africa and South America had been close together at the time of the Pennsylvanian period of the ice age.

The geosyncline concept was first developed by American geologists James Hall (1811–1898) and James Dwight Dana (1813–1895) in the mid 1800s. Dana was the first to use the term geosynclinal in reference to a gradually deepening and filling basin resulting from his concept of crustal contraction due to a cooling and contracting Earth. Even though the theory now stands discredited, Geosyncline is a term still occasionally used for a depression in the crust that has been caused by the accumulation of sedimentary rock strata deposited in a basin and subsequently compressed, deformed, and uplifted into a mountain range. Such a crustal depression shows features of volcanism such as rock formation from lava.

**JAMES DWIGHT DANA (1813-1895)** American geologist and mineralogist, and professor at Yale University from 1850

What Wegener argued was that as the continents moved, their edges encountered resistance and thus compressed and folded upwards to form mountains. He cited as examples the Sierra Nevada mountains on the Pacific coast of North America and the Andes mountains on the coast of South America. Wegener also suggested that India drifted northward into the Asian continent, thus forming the Himalayas. Observations in the later years confirmed several of Wegener's ideas, especially the work of the American geophysicist David Tressel Griggs (1911-1974), who showed that the mountain ranges on the Pacific coast were formed by the ocean floor plunging beneath the continents. In a dozen important papers published between 1934 and 1941, Griggs also proposed that thermal convection in deformable rocks in the earth's mantle is fundamentally responsible for the major physiographic features of the earth's surface.

## THE 'DRIFTERS'

The early adherents to Wegener's continental drift theory were called "Drifters". The Drifters visualized continents as sliding over the ocean floor. Most geophysicists at the time regarded this as impossible. Wegener eventually proposed a mechanism for continental drift whereby the centrifugal force created by the rotation of the earth would push the continents towards the equator. He argued that the original landmass formed near the South Pole and that the centrifugal force caused the proto-continent to break apart and the resultant continents to drift towards the equator. In 1929, about the time Wegener's ideas began to be dismissed, the British geologist Arthur Holmes (1890–1965) elaborated on one of Wegener's many hypotheses: the idea that the mantle undergoes thermal convection.

As a substance is heated, its density decreases and it rises to the surface until it cools and sinks again. In the case of

**OCEAN FLOOR FISSURE**

This metre-wide fissure is due to volcanic activity at a mid-ocean ridge. Mid-ocean ridges are formed when magma rises from inside the Earth through a crack created by spreading tectonic plates.

the earth's variously heated layers, the current caused by this repeated heating and cooling may be enough to move the continents. Holmes suggested that this thermal convection worked like a conveyor belt and that the upthrust of the pressure could break apart a continent and the convection currents could push the broken parts in opposite directions. This idea, too, received very little attention at the time.

In the following decades, a greater understanding of the ocean floor revived interest in Holmes' ideas. Discoveries of mid-oceanic ridges and the observation that arcs of islands and oceanic trenches occurred together and near the continental margins, suggested convection might indeed be at work. In 1947, a team of scientists led by the American geologist Maurice Ewing (1906–1974) utilizing the research vessel *Atlantis* of the Woods Hole Oceanographic Institution (the largest independent oceanographic research institution in the US), confirmed the existence of a rise in the central Atlantic Ocean, and found that the floor of the seabed beneath the layer of sediments consisted of basalt and not granite, which is the main constituent of continents. They also found that the oceanic crust was much thinner than continental crust.

# EARTH'S CURIOUS MAGNETISM

In the 1950s, geologists discovered that rocks crystallized from volcanic lava contained small quantities of magnetite, a strongly magnetic mineral. It was known that basalt, the iron-rich volcanic rock making up the ocean floor, contained magnetite (because it could locally distort compass readings). Geologists also found that rocks of similar ages found in different parts of the planet had the same magnetic characteristics. Rocks of different periods bore material differently aligned in the magnetic field, suggesting that the earth's magnetic field had reversed several times and the intervals between the reversed polarity had been extremely irregular.

These findings led to the new science of paleomagnetism, the study of the record of the earth's magnetic field. Scientists found a striped pattern of magnetic reversals in the crust of ocean basins, as evident in the alignment of the magnetite in seafloor basalt. As more and more of the seafloor was mapped, the magnetic variations revealed a zebra-like pattern. Alternating stripes of magnetically different rock were laid out in rows on either side of the mid-ocean ridge: one stripe with 'normal' polarity and the adjoining stripe with reversed polarity. The overall pattern, defined by these alternating bands, became known as magnetic striping.

The Canadian geologist Edward A. "Ted" Irving (b. 1927) was one of the key figures in the early studies of paleomagnetism and his work provided the first physical evidence of the theory of continental drift.

## SEAFLOOR SPREADING

An array of seismometers installed around the world in the 1960s to monitor nuclear testing revealed another startling geological phenomenon. It showed that earthquakes, volcanoes, and other active geologic features were for the most part aligned along distinct belts around the world, and those belts defined the edges of tectonic plates. This discovery, along with the several new findings around the same time, meant that there now was much more evidence to support the ideas of Wegener and Holmes. The major new advance in the theory of continental drift and thermal convection came in the shape of the theory of seafloor spreading.

Underneath the Earth's solid crust (lithosphere) is a malleable layer of heated rock called the asthenosphere, which is heated by the radioactive decay of elements such as uranium, thorium, and potassium. Because the radioactive source of heat is

**VOLCANIC GAS BUBBLES EMERGING THROUGH A CORAL REEF**

More evidence of seafloor spreading: volcanic gases rising through the seafloor off the coast of Volcano Island in Sangean, Indonesia.

A painting of the submerged volcanoes near Hawaii in the Pacific Ocean. These volcanoes have been lying under water since 30 million years.

deep within the mantle, the fluid asthenosphere circulates as convection currents below the lithosphere. This heated layer is the source of the lava we see in volcanoes, the source of heat that drives hot springs and geysers, and the source of the raw material that pushes up the mid-oceanic ridges and forms new ocean floor. Magma continuously wells upwards at the mid-oceanic ridges producing currents of magma flowing in opposite directions and thus generating the forces that pull the seafloor apart at the mid-oceanic ridges. As the ocean floor is spread apart, cracks appear in the middle of the ridges allowing the molten magma to surface and form the newest ocean floor. As the ocean floor moves away from the mid-oceanic ridge, it eventually comes into contact with a continental plate and is subducted, or slides underneath, the continental plate. Finally, the lithosphere is driven back into the asthenosphere where it returns to a heated state.

A major implication of seafloor spreading is that new crust was, and is now, being continually created along the oceanic ridges. This idea found great favor with some scientists, most notably the Australian geologist S. Warren Carey (1911–2002), who claimed that the shifting of the continents can be simply explained by the fact that the Earth had greatly increased in size since its formation. This so-called "Expanding Earth" hypothesis, however, could never be proven. The American geologist Bruce Charles Heezen (1924–1977), who led the team from Columbia University that mapped the Mid-Atlantic oceanic ridge during the 1950s, was an initial Carey supporter but later suggested that new oceanic crust formed at ridges on the oceanic floor.

The idea of a continuous re-creation of the ocean floor particularly intrigued Harry Hess (1906–1969), a Princeton University geologist and a Naval Reserve Rear Admiral, and Robert S. Dietz (1914–1995), a scientist with the US Coast and Geodetic Survey. Dietz and Hess coined the term seafloor spreading and argued that if the earth's crust was expanding along the oceanic ridges, it must be shrinking elsewhere. Dietz's 1961 paper and Hess's 1962 paper together suggested that new oceanic crust continuously spreads away from the ridges in a conveyor belt-like motion. Many millions of years later, the oceanic crust eventually descends into the oceanic trenches—very deep, narrow canyons along the rim of the Pacific Ocean basin. According to Hess, the Atlantic Ocean was expanding while the Pacific Ocean was shrinking. As old oceanic crust is consumed in the trenches, new magma rises

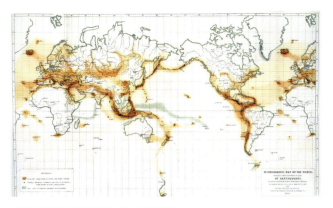

**SEISMOGRAPHIC WORLD MAP, 1857**

This map, centred on the Pacific and produced in 1857, shows the world's earthquake zones (orange). Areas of subsidence are blue, and volcanoes are marked as black dots. The seismic activity occurs along the active boundaries between the Earth's tectonic plates.

and erupts along the spreading ridges to form new crust. In effect, the ocean basins are perpetually being "recycled", with the creation of new crust and the destruction of old oceanic lithosphere occurring simultaneously.

## PLATE TECTONICS

The theories of continental drift and seafloor spreading were integrated in the 1960s in an overarching theory called plate tectonics. The British marine geologist Drummond Hoyle Matthews (1931–1997) was an important contributor to the theory of plate tectonics. His work, along with that of fellow Briton Frederick John Vine (b. 1939) and Canadian geologist Lawrence (Whitaker) Morley (b. 1920), showed how variations in the magnetic properties of rocks forming the ocean floor could be consistent with Harry Hess's 1962 theory of seafloor spreading. The

work of French geophysicist Xavier Le Pichon (b. 1937), British geophysicist Dan McKenzie (b. 1942) and American geophysicist Jason Morgan (b. 1935) was crucial in the development of the theory, as they had deducted from studies of magnetic polarity that the great tectonic plates moved relative to each other. The Deep Sea Drilling Project in the 1970s helped provide empirical evidence of many of these theories.

The basic concept of plate tectonics is that the earth's lithosphere exists as separate and distinct tectonic plates, which ride on the fluid-like asthenosphere. There are seven large plates and a dozen or more small plates. The location where two plates meet is called a plate boundary. Each plate is about 50 miles (80 km) thick and moves relative to another plate. The relative velocity between adjacent plates ranges from less than 7 cm per year to 13 cm per year, which is extremely high by geological standards.

## PLATE BOUNDARIES

Three types of plate boundaries exist, characterized by the way the plates move relative to each other. Transform boundaries occur where plates slide, or more accurately, grind past each other. The earthquake-prone San Andreas Fault in California is a transform boundary. Divergent boundaries occur where two plates pull apart from each other. Mid-oceanic ridges are divergent boundaries. Convergent boundaries have two plates sliding towards each other, commonly forming either a subduction zone (where a plate slides beneath another) or a continental collision (if the two plates consist of continental crust). Examples of this are the Andes mountains and the Japanese island arc.

**TECTONIC PLATE BOUNDARIES**
Artwork showing oceanic plates (gray) spreading from mid-ocean ridges (left), being destroyed at subduction zones (centre), and creating volcanoes at land and sea (right).

Plate boundaries, especially mid-oceanic ridges, are commonly the site of geological disturbances such as earthquakes and the source of geological activity that creates surface features such as mountains and volcanoes. The majority of the world's active volcanoes occur along plate boundaries. The Pacific Plate's Ring of Fire, a 40,000-km horseshoe-shaped series of oceanic trenches and volcanic belts, has the densest concentration of volcanoes. The Ring of Fire has 452 volcanoes, which make up over 75 per cent of the world's active and dormant volcanoes.

## MID-OCEANIC RIDGES

It is at the mid-ocean ridges, which rise 3000 meters from the ocean floor, that the seafloor spreads and new crust is created. All the mid-ocean ranges of the world are connected and form the longest mountain range in the world. The continuous mountain range is 40,400 miles (65,000 km) long and the total length of the system is 49,700 miles (80,000 km). The mapping of the seafloor also revealed that these huge ranges have a deep trench that bisects the length of the ridges and in places is more than 2000 meters deep. The mountain ranges also typically have a valley known as a rift running along their spine.

## DEEP SEA TRENCHES

The deepest waters are found in the oceanic trenches, which plunge as deep as 35,000 feet (10,668 meters) below the ocean surface. These trenches are usually long and narrow, and are often associated with and lie parallel to large continental mountain ranges. Like the mid-oceanic ridges, the trenches are seismically active, but unlike the ridges they have low levels of heat flow. Scientists have determined that the youngest regions of the ocean floor lie along the mid-oceanic ridges, and that the age of the ocean floor increases as the distance from the ridges increases. The oldest seafloor often ends in the deep-sea trenches. A major trench is the Java-Sumatra trench. Chains

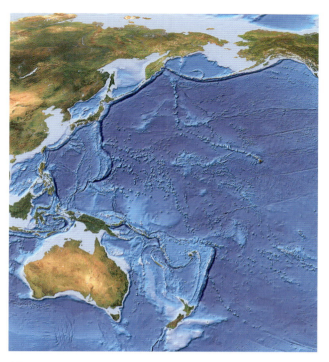

**OCEAN TRENCHES**
Computer model of the Pacific Ocean, based on satellite data. Bare land is brown, vegetation is green and snow and ice are white. The Mariana Trench (dark blue crescent, top) is the deepest ocean trench on earth.

of islands are found throughout the oceans and especially in the western Pacific margins. These "island arcs" are usually situated along deep sea trenches and on the continental side of the trench. The Japanese, Indonesian, Philippine, and the Solomon islands are examples of such arcs.

## EARTHQUAKES

Most earthquakes occur along or near plate boundaries. They are sudden releases of energy caused by a fault, created when a rock surface slips against another, fracturing the plane. Such faults occur regularly and are concentrated in certain fault zones, such as in the Japanese island arc. Earthquakes are caused mostly by rupture of geological faults, but also

**TSUNAMI OR 'TIDAL WAVE'**

Illustration depicting a tsunami, with an erupting volcano in the background. Tsunamis are caused by earthquakes. Over deep ocean, the mass of uplifted water appears as a low and fast-moving disturbance. Over shallow coastal seas, the velocity drops but the amplitude increases, producing an enormous wave capable of immense destruction. The largest recorded tsunami was 85 meters high, and struck the Japanese Ryukyu Islands on April 24, 1771.

by volcanic activity, landslides, mine blasts, and nuclear tests. An earthquake's point of initial rupture is called its focus or hypocenter. The term epicenter refers to the point at ground level directly above the hypocenter. When a large earthquake epicenter is located offshore, the seabed sometimes suffers sufficient displacement to cause a tsunami, a seismic sea wave. The intensity of earthquakes is measured on the Richter scale, a base-10 logarithmic scale. Earthquakes with magnitude above 7 on the Richter scale can cause huge devastation.

## MINERALS

Simultaneous with the revolution in the understanding of geological processes, a great surge took place in the exploration and exploitation of resources contained in the earth. All the metals and most of the materials of use to man, other than cloth and wood, are obtained from the earth's crust. A mineral is a naturally occurring solid that has a characteristic chemical composition, a crystalline atomic structure, and specific physical properties. Minerals range in composition from pure elements and simple salts to very complex silicates with thousands of known forms. Minerals are extracted from rocks, sea or river beds, and from beneath the ground. Most minerals are inorganic, i.e., contain no carbon. There are currently more than 4000 known minerals according to the International Mineralogical Association. Of these, 100 can be called "common", 50 as "occasional", and the rest as "rare" to "extremely rare".

Rocks are sometimes referred to as aggregates of minerals, but some rocks like limestone are composed entirely of one mineral. To the geologist, clay and loose sand are as true rocks as the hardest of granite or sandstone. Rocks are formed by solidification from magma (igneous rocks), deposits (sedimentary rocks), and by intense pressure and temperature on an igneous or sedimentary rock (metamorphic rocks).

Most of the rock-forming minerals are silicates, but the great majority of metals are found in ores that contain no silicates. Except iron, the metals used for industrial and daily purposes

**GOLDMINING IN CALIFORNIA, 1871**

Gold is one of the rarest minerals and exists in the earth's crust as a trace among a mixture of other minerals such as quartz, pyrite and carbonates.

occur in the earth's crust in very low amounts. Their share could be as low as 10-16 parts per million (ppm). Metals can be mined only when they have been enriched by geological process to a specific concentration. In most ore deposits, the metal is intermixed with minerals like quartz that have no economic value. For example, gold veins often are made up of large amounts of quartz and carbonates, with some pyrite and a little gold. These unwanted minerals in the ore are called gangue and are discarded during the processing.

## KINDS OF MINERALS

Minerals can be divided into metallic and non-metallic types. Non-metallic minerals range from common clay to diamonds, and form the basic building blocks of all structures. In construction, cement is made from a mixture of limestone and clay, glass from quartz sand and quartzite, plaster from gypsum, insulatory material from mica and vermiculite, and granite, marble, etc., serve as readymade building material.

Ceramics and refractory materials, which retain their strength at high temperatures, are made from oxides such as of aluminium (alumina), silicon (silica), magnesium (magnesia) and calcium (lime). Fireclays, commonly found with coal deposits, are also widely used in the manufacture of refractory and ceramic materials. Refractory materials are used extensively in metallurgy, along with other heat treatment operations such as glass melting. Traditional ceramic raw materials include clay minerals such as kaolinite, while modern ceramic materials, which are classified as advanced ceramics, include silicon carbide and tungsten carbide.

Apart from metals, precious and semi-precious stones are also used for jewelry and decorative purposes. Diamonds are widely mined in Namibia and South Africa, rubies and sapphires in Myanmar, Thailand and Sri Lanka, emeralds in Brazil and Zimbabwe, topaz (which can be yellow, pink or blue) in the Ural mountains and Brazil, zircon (a pale colored gem) in Sri Lanka and Myanmar, the purple gem amethyst in Brazil and India, and the variously colored garnet in South Africa and India.

**MOUNTAINTOP REMOVAL MINING**

A levelled section of mined land in the Appalachian Mountains, USA, after a large section of the mountain (called the overburden) has been blasted to access a seam of coal lying below it.

# HISTORY OF MINING

Flint, which could be broken into sharp-edged pieces that were useful as scrapers, knives, and arrowheads, was the first mineral to be mined. During the Neolithic Period, or New Stone Age (8000-2000 BC), shafts up to 330 feet (100 meters) deep were sunk in soft chalk deposits in France and Britain to extract the flint pebbles found there. Other minerals, such as red ochre and the copper mineral malachite, were used as pigments. The oldest known underground mine in the world was sunk more than 40,000 years ago at Bomvu Ridge in the Ngwenya Mountains, Swaziland, to mine ochre used in burial ceremonies and as body coloring.

According to historians, the Egyptians were mining copper on the Sinai Peninsula as long ago as 3000 BC, although some

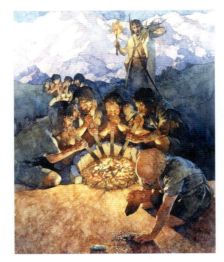

**PREHISTORIC MEN SMELTING COPPER**
After iron, copper was the most useful metal to be extracted, and was being mined in 3000 BC.

bronze (copper alloyed with tin) is dated as early as 3700 BC. Earliest iron is dated to 2800 BC and lead was being produced in 2500 BC (it is found in the ancient ruins of Troy). One of the earliest uses of quarried stone for building purposes was

in the construction (2600 BC) of the great pyramids in Egypt, the largest of which (Khufu) is 236 metres along the base sides and contains approximately 2.3 million blocks of two types of limestone and red granite. The limestone is believed to have been quarried from across the Nile.

One of the most complete early accounts of mining methods in Europe is by the German scholar Georgius Agricola in his *De re metallica* (1556). He describes detailed methods of driving shafts and tunnels. Great progress in mining was made when the secret of black powder reached the West, probably from China in the late Middle Ages. This was replaced as an explosive in the mid 19th century with dynamite.

Water inflow was a major problem in underground mining until James Watt (1736–1819) invented the steam engine in the 18th century. After that, steam-driven pumps could be used to remove water from the deep mines. Early lighting systems were of the open-flame type, consisting of candles or oil-wick lamps. This often reacted with the methane in the mines to cause explosions. Humphry Davy's (1778–1829) invention of miners' safety lamps in the early 19th century solved this problem. In the 1930s, battery-powered cap lamps began to be used. Technology has now developed to the point where gold is mined at depths of 4000 meters, and mines have been excavated to more than 700 meters.

**SEARCHING FOR IRON ORE IN CHINA** A painting depicting the search for iron in ancient China. China has today the most extensive network of mines and produces the bulk of the world's coal.

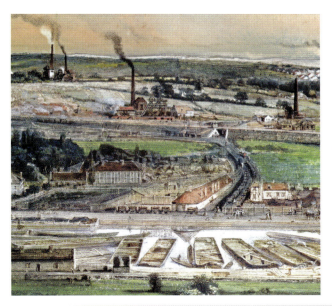

**COAL FIELD, SAONE-ET-LOIRE**

A watercolor on paper showing the industrial landscape in the Blanzy coal field at Saone-et-Loire in France in 1860.

**19TH-CENTURY MINE WORKERS**

An 1868 artwork depicting coal miners going down a mineshaft. The shield above the miners' heads protects them from falling stones.

## COAL MINING

Coal is the most abundant fossil fuel on Earth. It was the main energy source that fueled the Industrial Revolution of the 18th and 19th centuries. Since the mid-20th century, coal has yielded its place to petroleum and natural gas as the principal energy supplier of the world. The mining of coal from surface and underground deposits today is a highly productive, mechanized operation. Global coal production is expected to reach 7000 million tons per year in 2030, with China accounting for most of this increase. At current production levels, proven coal reserves are estimated to last 147 years.

Archeological evidence suggests that coal was burned in funeral pyres during the Bronze Age, 3000 to 4000 years ago, in Wales. Aristotle mentions coal ("combustible bodies") in his *Meteorologica*, and his pupil Theophrastus also records its use. The main techniques of underground coal mining, however, were developed primarily in Britain in the late 18th century.

International trade in coal expanded exponentially when coal-fed steam engines were built for the railways and steamships between 1810 and 1840. Coal was cheaper and much more efficient than wood fuel for most steam engines. As central and Northern England contained an abundance of coal, many mines were situated in these areas as well as in Wales and Scotland.

Coal deposits are found in sedimentary rock basins, where they appear as successive layers, or seams, sandwiched between strata of sandstone and shale. There are more than 2000 coal-bearing sedimentary basins distributed around the world. World coal resources, or the total amount of coal known to exist, are approximately 11 trillion tons. The distribution of the estimated coal resources of the world is approximately as follows: Europe (including Russia and the former Soviet republics) 49 per cent; North America 29 per cent; Asia 14 per cent; Australia 6 per cent; and Africa and South America 1 per cent each. Distinct from coal resources are coal reserves, which are only those resources that are technically and economically minable at a particular time.

**MARAS SALT MINE, PERU**

Salts such as chlorides and sulfates of potassium and magnesium are mined from large, shallow ponds exposed to the sun.

## MARINE MINING

Although the sea is a major storehouse of minerals, it has been little exploited. Marine mining, however, is one of the major areas of growth in the worldwide exploration of mineral resources. In the context of mining, the sea can be divided into three regions: seawater, beaches and continental shelves, and the seafloor.

Seawater contains by weight an average of 3.5 per cent dissolved solids. The most important constituents, in decreasing order, are chlorides, sodium, sulfates, magnesium, calcium, potassium, bromine, and bicarbonates. In addition to the oceans, minerals are also recovered from the waters of inland salt seas, the Dead Sea and the Great Salt Lake being two notable examples. While seawater is an important source of magnesium, by far the most common minerals extracted from seawater are salts, especially common table salt (sodium chloride, NaCl), the chlorides of potassium and magnesium, and the sulfates of potassium and magnesium. These minerals are mined by evaporation, very often in large, shallow ponds with energy being supplied by the Sun.

Although micas, feldspars and other silicates, and quartz form the bulk of the material on most beaches, considerable quantities of valuable minerals such as columbite, magnetite, ilmenite, rutile, and zircon are also commonly found. All these are classified as heavy minerals and are generally resistant to chemical weathering and mechanical erosion. Less commonly found in minable concentrations are gold, diamonds, cassiterite, scheelite, wolframite, monazite, and platinum. For the mining of beach deposits above sea level, conventional techniques of surface mining are used.

The floors of the great ocean basins consist of gently rolling hills. The dominant seafloor sediments are oozes and clays, and the most important mineral deposits known (but not yet exploited) are phosphorite and manganese nodules. From an economic standpoint, the manganese nodules (actually concretions of manganese dioxide) are more important, and are found in largest concentration in the eastern Pacific between Hawaii and Central America.

## QUARRYING

It has been estimated that more than two-thirds of the world's yearly mineral production is extracted by surface mining. There are several types of surface mining, but the three most common are open-pit mining, strip mining, and quarrying. Among these, quarrying is mainly used to extract building stone

**THE SLATE QUARRIES AT PENRHYN, 1832**

The United Kingdom has rich mineral reserves and a long tradition of mining. This quarrying mine of the building stone, slate, existed in Wales.

such as granite, limestone, sandstone, marble, slate, gneiss, and serpentine. All natural stone used for structural support, curtain walls, veneer, floor tiles, roofing, or strictly ornamental purposes is called building stone, and building stone that has been cut and finished for predetermined uses in building construction and monuments is known as dimension stone.

## PRECIOUS RESOURCES

By the end of the 20th century, applications of the science of geology had become focused on the need to exploit the earth's resources in a sustainable fashion. New technology helped make use of materials that were earlier considered of little value or rare and unreachable. The great surge in the use of computers was possible because of technology that worked with newer silicates and alloys. So extensive has been the consumption of metals like gold, copper and silver in the computer hardware industry that computer parts have become

a virtual mine of these metals. It wouldn't be off the mark to say that the recycling industry is a parallel mining industry in today's times and that computers, automobiles, and electronic machines have become the new mineral 'ores'.

The surge in the exploration of ocean floors, river systems, and land surface features has exponentially enlarged the prospective scope for mineral exploration. But the major concern on the ground remains that precious minerals such as fossil fuels are not completely used up within a short time. There is also a growing realization that the polluting waste generated in the manufacture and use of industrial products has to be reduced. As a consequence, the focus of the energy industry has shifted to making use of renewable and non-polluting natural resources. In the next chapter, we'll see how this seismic shift (to borrow a geological term) is playing out in the field of the Earth sciences.

**NORTH SEA OIL RIG**

A computer-generated artwork of an oil drilling platform in the North Sea, off the Scottish Coast, on a stormy night.

# Chapter 4
# A FRAGILE PLANET

Smog is the fog-like haze created by the smoke from vehicles, factories, and other kinds of chemical combustion. It is the most visible form of air pollution in cities and can cause severe heart and lung diseases.

**FACING PAGE** A digitally generated picture, suggesting the consequences of global warming, shows a man using a kayak boat to get past a flooded Times Square in New York.

At the turn of the 21st century, realization set in that the exploitation of the earth's resources could not remain unbridled. Several decades of exploration for minerals had brought an increasing awareness of the earth and its environment. Improved techniques of map-making, the data generated by satellites, a range of geophysical surveys, and new technologies that made it possible to reach hitherto inaccessible places such as ocean floors and outer space, completely changed the existing worldview. Humans were shown to have affected the earth much more substantially, and adversely, than had previously been thought. This chapter examines three areas where the major work on earth studies is taking place. These areas are: new information technologies, oceanography, and global warming.

## GEOGRAPHICAL INFORMATION SYSTEM

Any information system that can store and edit geographic data can be classified as a Geographical Information System (GIS). In a more general sense, GIS applications are tools that allow users to create interactive queries (user-created searches), analyze spatial information, and edit data and maps.

The world's first operational GIS was developed in 1962 in Ottawa, Canada, by Roger Tomlinson (b. 1933) for the Department of Forestry and Rural Development. It was called the Canada Geographic Information System (CGIS) and it lasted into the 1990s, having built the largest digital land resource database in Canada.

By the early 1980s, Alabama-based M&S Computing (later Intergraph), California-based Environmental Systems Research Institute (ESRI) and Canada-based CARIS (Computer Aided Resource Information System) emerged as commercial vendors of GIS software. Public access to geographic information is mostly dominated by online resources such as Google Earth and interactive web mapping. The Open Geospatial Consortium (OGC), an international industry consortium of 334 companies, government agencies and universities, is trying to build a consensus to develop publicly available geoprocessing specifications.

A GIS can convert digital information into a map form that can be integrated with other maps. For example, satellite images can be analyzed to produce a map-like layer of digital information

**ENVISAT-1 SATELLITE IN ORBIT**

The first satellite of Europe's Polar-orbiting Earth Observation Mission (POEM), Envisat-1 monitors the earth's temperature, moisture, and pressure.

allowing for better understanding of terrestrial processes and better management of human activities to maintain economic growth and environmental quality.

GIS is also being explored for its ability to track and model the progress of humans throughout their daily routines. A concrete example of progress in this area is the recent release of time-specific population data by the US Census. In this data set, the populations of cities are shown for daytime and evening hours highlighting the pattern of concentration and dispersion generated by North American commuting patterns.

## REMOTE SENSING

The use of imaging sensor technologies that are located in space is referred to as remote sensing. One example of this is Radar (radio detection and ranging) remote sensing, a

**ICE-PENETRATING RADAR SYSTEM**

Glaciologists measuring the thickness of the Taylor Glacier in Antarctica's Taylor Valley in November 2005 using a ground-penetrating radar system.

about vegetative covers. Likewise, census or hydrological data can be converted into a map. GIS technology gives researchers the ability to examine the variations in Earth processes over days, months, and years. This enables researchers to detect regional differences in, for example, the lag between a decline in rainfall and its effect on vegetation.

GIS data represents real world objects (roads, land use, elevation) with digital data. Real world objects can be divided into two abstractions: discrete objects (a house) and continuous fields (rainfall amount or elevation).

The condition of the Earth's surface, atmosphere, and subsurface can be examined by feeding satellite data into a GIS. GIS technology gives researchers the ability to examine the variations in Earth processes over days, months, and years. This enables researchers to detect regional differences in, for example, the lag between a decline in rainfall and its effect on vegetation. GIS and related technology helps in the management and analysis of these large volumes of data,

system that uses electromagnetic waves to identify the range, altitude, direction, or speed of both moving and fixed objects such as aircraft, ships, motor vehicles, weather formations, and terrain.

Remote sensing makes it possible to collect data on dangerous or inaccessible areas, and its applications include monitoring deforestation in regions such as the Amazon Basin, the effects of climate change on glaciers, and the measurement of coastal and ocean depths. Measuring instruments aboard satellites such as Landsat, the Nimbus, and more recent missions such as RADARSAT and UARS, provided global measurements of various data for civil, research, and military purposes. Remote sensing data is processed and analyzed with computer software, known as a remote sensing application. A large number of commercial and open source applications exist to process remote sensing data.

## GROUND-PENETRATING RADAR

The ground-penetrating radar (GPR) uses radar pulses to image objects below the surface. GPR can be used to study rocks, soil, ice, underground water, and structures such as roads and tunnels. It can detect objects, changes in material, and voids and cracks. The depth range of GPR is, however, adversely affected by any electrical conductivity in the ground. Optimal depth penetration is achieved in ice, where objects several hundred feet below can be imaged. Penetration in dry sandy soils or massive dry materials such as granite, limestone, and concrete, can reach up to 49 feet (15 meters) inside the surface. In moist and/or clay-laden soils, penetration is sometimes only a few inches.

In Earth sciences, GPR is used to study bedrock, soils, groundwater, and ice. Engineering applications include locating buried structures and utility lines. GPR also helps identify sites of hazardous waste and contamination. In archaeology, it is

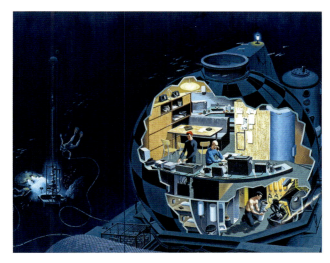

**CONSHELF-III**

Conshelf, or Continental Shelf Station, were three undersea research stations built in the 1960s. Conshelf-III had six divers living 100 meters under the Mediterranean Sea for three weeks in 1965.

used for mapping archaeological features and cemeteries. Military uses include detection of mines and tunnels.

## OCEANOGRAPHY

Oceanography, also called oceanology or marine science, is the branch of Earth science that studies the ocean. It is the base science of Ocean engineering, which involves designing and building oil platforms, ships, harbors, and other structures that allow the safe use of the ocean.

American explorer Matthew Fontaine Maury's (1806–1873) *Physical Geography of the Sea* (1855) was the first textbook of oceanography. Oceanography began as a quantifiable science in 1871, when under the recommendations of the Royal Society of London, the British government sponsored an expedition to scientifically explore the world's oceans. The expedition, called the Challenger expedition (1872-1876), was launched in 1872 by two Scottish explorers, Charles Wyville Thompson (1830–1882) and Sir John Murray (1841–1914). The results thereof were published in 50 volumes covering biological, physical and geological aspects, including the discovery of 417 new marine species. Other European and American nations also sent out scientific expeditions (as did private individuals and institutions). The first purpose-built oceanographic ship, the Albatross, was built in 1882. In 1910, a four-month North Atlantic expedition headed by Sir John Murray and Johan Hjort (1869–1948) was at that time the most ambitious research oceanographic and marine zoological project, and led to the classic 1912 book, *The Depths of the Ocean*.

Oceanographic institutes dedicated to the study of oceanography were founded. In the United States, these included the Scripps Institution of Oceanography in 1892, Woods Hole Oceanographic Institution in 1930, Lamont-Doherty Earth Observatory at Columbia University, and the School of Oceanography at University of Washington. In Britain, the major research institution, National Oceanography Centre, Southampton, is the successor to the Institute of Oceanography. In Australia, CSIRO Marine and Atmospheric Research, known as CMAR, is a leading center. In 1921, the International Hydrographic Bureau was formed in Monaco.

In 1893, Fridtjof Nansen (1861–1930) allowed his ship *Fram* to be frozen in the Arctic ice. As a result, he was able to obtain oceanographic as well as meteorological data. The first international organization of oceanography was created in 1902 as the International Council for the Exploration of the Sea. The first acoustic measurement of sea depth was made in 1914. Between 1925 and 1927, the "Meteor" expedition, surveying the Mid-Atlantic ridge, gathered 70,000 ocean depth measurements using an echo sounder. The Great Global Rift, running along the Mid-Atlantic Ridge, was discovered by Maurice Ewing (1906–1974) and Bruce Heezen (1924–1977) in 1953. The mountain range under the Arctic was found in 1954 by the Arctic Institute of the USSR.

From the 1970s there has been much emphasis on the application of computers to oceanography to allow numerical predictions of ocean conditions and as a part of overall environmental change prediction. An oceanographic buoy array was established in the Pacific to allow prediction of El Niño events.

**OIL RIG**

An offshore oil drilling platform, or oil rig, operated by the US company Shell extracts oil and gas from the ocean floor.

**CLIMATE CHANGE ACTIVIST AND POLITICIAN AL GORE**

Former US Vice President Al Gore shared the 2007 Nobel Peace Prize with the U.N. Intergovernmental Panel on Climate Change for raising government and corporate concern on effects of climate change.

**ILULISSAT KANGERLUA GLACIER MELTS**

An iceberg, part of the Ilulissat Kangerlua Glacier in Greenland, melts due to the effect of a sharp increase in temperature in recent years.

## DEEP SEA DRILLING PROJECT

The Deep Sea Drilling Project (DSDP), established in June 1966 as a collaborative program of several nations, operated the *Glomar Challenger* ship in drilling operations in the Atlantic, Pacific, and Indian Oceans, as well as in the Mediterranean and Red Seas. In 1983, the DSDP was restructured as the Ocean Drilling Program (ODP), which in turn became the Integrated Ocean Drilling Program (IODP) in 2003. The IOPD was conceived as a 10-year Earth science and research program. The Deep Sea Drilling Project and its later variations generated a great deal of new information about the Earth's dynamic nature: tectonic processes, ocean circulation, climate change, continental rift, and ocean basin formation. Today, scientific ocean drilling continues to provide a powerful tool to study the critical processes related to Earth's short-term change and long-term variability, and is a key supplier of data on climate change.

## THE LAW OF THE SEA

The 1982 United Nations Convention on the Law of the Sea was an international agreement that resulted from the third United Nations Conference on the Law of the Sea (UNCLOS III). The Convention defines the rights and responsibilities of nations in their use of the world's oceans and establishes guidelines for protecting the marine environment and managing marine resources. Currently, the European Union and 156 countries have ratified the UN Convention on the Law of the Sea.

The most significant provision introduced by the convention is the setting of territorial limits. Every nation that has a coastline is free to set laws, regulate use, and exploit any resource up to 12 nautical miles from its defined baseline. Beyond that, a further 12 nautical miles (i.e., 24 nautical miles from the baseline) constitute the contiguous zone, in which a state can

**MANGANESE NODULES**

The floor of the northeast Atlantic Ocean, showing a large concentration of manganese nodules, minerals that are a major focus of deep-sea mining. These nodules form when manganese and iron minerals are deposited on the surface of a tiny object, such as a piece of bone.

enforce laws in four specific areas: pollution, taxation, customs, and immigration. Other significant measures introduced by the Convention include defining the status of archipelagos and transit regimes, creating exclusive economic zones (EEZs) in seawaters as an incentive to industry, defining continental shelf jurisdiction, regulating deep seabed mining, and framing rules for settling disputes concerning the sea.

## INTERNATIONAL SEABED AUTHORITY

The International Seabed Authority (ISA) is an intergovernmental body based in Kingston, Jamaica, that was established in 1994 to organize and control all mineral-related activities in the international seabed area beyond the limits of national jurisdiction. However, contrary to early hopes that seabed mining would generate extensive revenues for both the exploiting countries and the ISA, no technology has yet been developed for gathering deep-sea minerals at costs that can compete with land-based mines. Moreover, as the US has not ratified the Law of the Sea, and is therefore not a member of the ISA, it makes it quite difficult to organize concerted international efforts to invest in deep-sea mining.

In recent years, interest in deep-sea mining, especially with regard to ferromanganese crusts and polymetallic sulphides, has picked up. Ocean floors and seabeds are rich sources of such metals as copper, cobalt, zinc, nickel, lead, and gold. Several firms are now operating in the waters of Papua New Guinea, Fiji, and Tonga. Papua New Guinea became the first country in the world to grant commercial exploration licenses when it signed a contract with the Canada-based firm Nautilus Minerals in 1997, giving the company rights over a major part of the country's massive seafloor sulphide deposits.

## GLOBAL WARMING

The greenhouse effect refers to the process by which the absorption and emission of infrared radiation by atmospheric gases leads to a warming of the earth's lower atmosphere and surface. Certain gases have a higher tendency to absorb and emit such radiation, and many of them are common in the earth's atmosphere, the commonest being water vapor, carbon dioxide, methane, nitrous oxide, and the group of gases known as chlorofluorocarbons (CFCs). Since the industrial revolution,

**BLEACHED CORAL**

This coral, off the Maldive Islands in the Indian Ocean, has been bleached due to the loss of symbiotic algae. The bleaching involves the loss of a vital pigment that protects the coral from radiation. Adverse changes in the marine environment, especially sea pollution and rising temperatures, are considered the main reason for the disappearance of the algae.

the amount of greenhouse gases in the atmosphere has increased enormously, raising the temperature of land, air, and water, a phenomenon referred to as global warming.

The greenhouse effect was discovered as far back as the 19th century by the French mathematician and physicist Joseph Fourier (1768–1830) in 1824, and it was first investigated quantitatively by the Swedish scientist Svante Arrhenius (1859–1927) in 1896. But it was only in the late 20th century that the role of greenhouse gases in the continual increase in temperature of the atmosphere, land, and oceans was realized.

The latest report (2007) of the United Nation's Inter Governmental Panel on Climate Change (IPCC) foresees a rise in global temperature of around 2.0 to 11.5°F (1.1 to 6.4°C) during the 21st century. This could cause sea levels to rise and could change the amount and pattern of rainfall, leading to more deserts in subtropical regions. Other likely effects include shrinkage of the Arctic ice shelf, depletion of Amazon rainforests, increase in the intensity of extreme weather events, reduced agricultural yields, glacier retreat, extinctions of species, and changes in the range and spread of disease-causing organisms. In 2007, The IPCC shared the 2007 Nobel Peace Prize with former US Vice-President Al Gore.

Although the reality and projections of climate change are routinely dismissed by industry, there is now scientific consensus that the increase in greenhouse gases has been caused by human activity since the start of the industrial era. The atmospheric concentrations of $CO_2$ and methane have increased by 36 per cent and 148 per cent respectively since the mid 1700s. The principal international effort to reduce greenhouse gas emissions has been through the Kyoto Protocol (1997), which, however, expires in 2012. Another important initiative to combat climate change has been an emissions

**GREENHOUSE EFFECT**

The left bottle contains ordinary air and the right bottle contains carbon dioxide ($CO_2$). Sunlight warms the piece of brown card, which corresponds to the earth's surface, and the card radiates this energy as infrared rays, which are trapped in the bottles. The temperature in the bottle containing $CO_2$ rises more, demonstrating the warming, or greenhouse, effect of carbon dioxide.

trading scheme that uses 'carbon credit' to encourage businesses to cut down greenhouse emissions. Any financial loss involved in reducing emissions is met by a carbon credit that shows up in the balance sheet as any other production cost. Also, a developed country can 'sponsor' a greenhouse gas reduction project in a developing country, where the cost of such reduction is usually much lower. The developed country would be given credits for meeting its emission reduction targets, while the developing country would benefit from the capital investment and the clean technology.

Thus, the 21st century stands at a key juncture in the history of man's relationship with the earth. It is estimated that by 2050 the world's population will have crossed 9 billion and 7 billion people will be living in cities. The lack of clean drinking water has already become a major problem in the developing world, showing what an unprecedented population pressure can do to the most abundant natural resource. The consequences may seem dire but the solution is within our grasp, if only we accept that the earth belongs to all who live on it.

# INDEX

# PICTURE CREDITS